2 What time is it? Write the time below each clock.

5 points per question

(1)   (2)   (3)   (4)

(   )   (   )   (   )   (   )

3 How long is each piece of tape? Write the length under each ruler.

5 points per question

(1)

1 inch

(   )

(2)

1 inch

(   )

4 Count the money in each row and write the amount on the right.

3 points per question

(1)   (   )

(2)   (   )

(3)   (   )

(4)   (   )

(5)   (   )

Remember this stuff? Good!

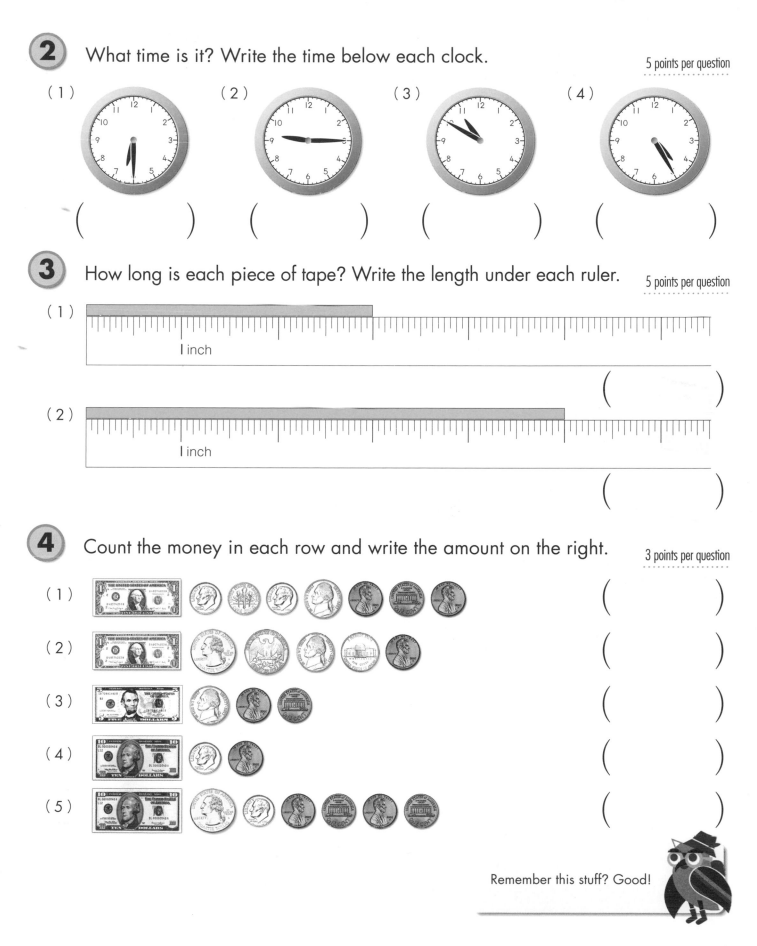

1 Fill in the missing number in each box on the number lines below.

2 points per box

(1)

9,910 9,920 9,930 ↓ 9,950 ↓ 9,970 9,980 ↓ 10,000

(2)

4,995 ↓ 4,997 ↓ 4,999 ↓ 5,001 5,002 ↓ 5,004 5,005

2 Write the appropriate numbers below.

3 points per question

(1) The number that is 1 more than 6,000.

(   )

(2) The number that is 1 less than 10,000.

(   )

(3) The number that is 1 more than 5,010.

(   )

(4) The number that is 1 less than 5,010.

(   )

(5) The number that is 1 more than 8,999.

(   )

(6) The number that is 1 less than 7,900.

(   )

3 Circle the larger number in each pair of numbers below.

4 points per question

(1) 7,900   9,700

(2) 5,430   5,340

(3) 1,061   1,016

(4) 6,767   6,677

(5) 4,331   4,333

(6) 8,080   8,108

KUMON MATH WORKBOOKS

Grade 3

Geometry & Measurement

Table of Contents

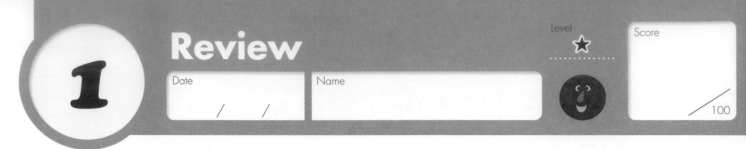

Review

1

Level ☆

Score

/100

Date / /

Name

1 What number is represented in each of the questions below?

5 points per question

(1)

100 100 100 100 100 100 100 100 100 100

100 100 100 100 10 10 10 10 10 10 | | | | |

(   )

(2)

1000 1000
1000 1000
1000

100 100 100 100   10 10   1 1 1

100 100 100   10 10

(   )

(3) The number you get from adding **7** hundreds, **9** tens and **2** ones.

(   )

(4) The number you get from adding **8** hundreds and **1** one.

(   )

(5) The number you get from adding **6** thousands, **3** hundreds, and **4** tens.

(   )

(6) The number you get from adding **10** thousands.

(   )

(7) The number you get from adding **23** tens.

(   )

(8) The number you get from adding **90** hundreds.

(   )

(9) The number you get from adding **86** hundreds.

(   )

(10) The number that has a **2** in the thousands place, a **5** in the hundreds place, a **3** in the tens place, and a **0** in the ones place.

(   )

(11) The number that has a **9** in the thousands place, a **0** in the hundreds place, a **0** in the tens place, and an **8** in the ones place.

(   )

4 Draw the long hand on the face of each clock to match the time above.

5 points per question

(1) (7:30) (2) (11:45) (3) (1:55)

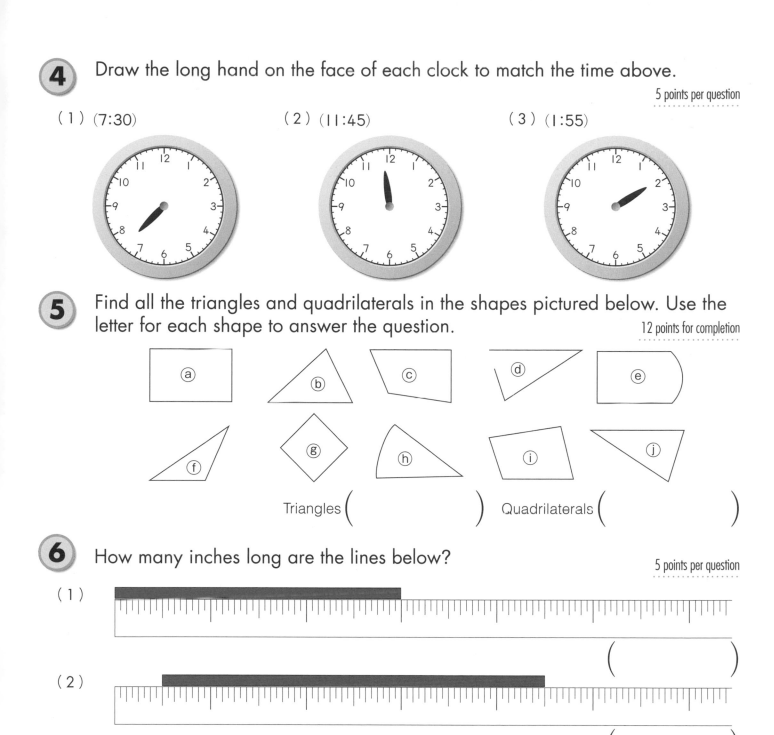

5 Find all the triangles and quadrilaterals in the shapes pictured below. Use the letter for each shape to answer the question.

12 points for completion

Triangles ( ) Quadrilaterals ( )

6 How many inches long are the lines below?

5 points per question

(1)

( )

(2)

( )

(3)

( )

Now it's time for something new!

3
Large Numbers

Level ★★

Date / /

Name

Score
/100

1 How many sheets of paper are pictured in each box below? Each bundle has 1,000 sheets.

6 points per question

(1)

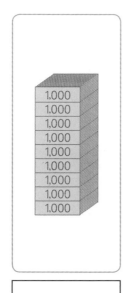

[　　　　　] sheets of paper

(2)

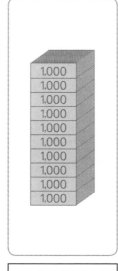

[10,000] sheets of paper

(3)

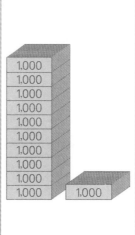

[　　　　　] sheets of paper

(4)

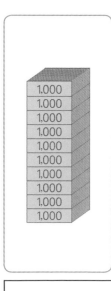

[　　　　　] sheets of paper

(5)

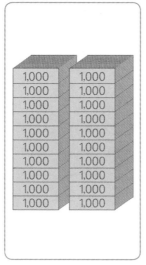

[　　　　　] sheets of paper

(6)

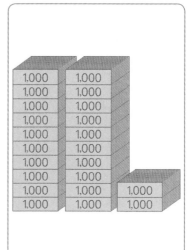

[　　　　　] sheets of paper

2 How many sheets of paper are pictured in each box below?

6 points per question

(1)

1,000
1,000
1,000
10,000 1,000
10,000 1,000 100
10 10 10 10 1 1 1

_____ sheets of paper

(2)

1,000
1,000
1,000
10,000 1,000
10,000 1,000 100
10,000 1,000 100 10
1 1 1 1 1

_____ sheets of paper

(3)

10,000
10,000
10,000 100 10
10,000 100 10 1
1 1 1

_____ sheets of paper

(4)

10,000
10,000
10,000
10,000
10,000 1,000
1 1 1 1 1 1

_____ sheets of paper

3 Write the appropriate numbers below.

8 points per question

(1) The number you get from adding 7 ten-thousands, 9 thousands, 4 hundreds, 6 tens and 2 ones. (   )

(2) The number you get from adding 5 ten-thousands, 2 thousands, 8 hundreds and 1 one. (   )

(3) The number you get from adding 4 ten-thousands and 5 thousands. (   )

(4) The number you get from adding 9 ten-thousands and 8 hundreds. (   )

(5) The number you get from adding 3 ten-thousands and 7 tens. (   )

Are you getting used to large numbers?
Good!

Large Numbers

4

Date   /   /

Name

Level ☆☆

Score   /100

---

1 Write the appropriate number in each box below.

5 points per question

(1) The number that is ten times one thousand is ten thousand and is written ⟨10,000⟩.

(2) The number that is ten times ten thousand is one hundred thousand and is written ⟨   ⟩.

(3) The number that is ten times one hundred thousand is one million and is written ⟨   ⟩.

(4) The number that is ten times one million is ten million and is written ⟨   ⟩.

---

2 Answer the following questions about the number 26,345,791.

4 points per question

(1)

place	place	place	ten-thousands place	thousands place	hundreds place	tens place	ones place
2	6	3	4	5	7	9	1

(2) Reading from the thousands place to the left, we see the ⟨   ⟩ place, the ⟨   ⟩ place, the ⟨   ⟩ place and the ⟨   ⟩ place.

(3) The **6** in 26,345,791 is in the ⟨   ⟩ place.

(4) The **2** in 26,345,791 is in the ⟨   ⟩ place.

(5) The **3** in 26,345,791 is in the ⟨   ⟩ place.

3 Write the appropriate numbers below.

5 points per question

(1) Write the number that has a **4** in the ten-millions place, a **3** in the millions place, a **5** in the hundred-thousands place, a **7** in the ten-thousands place and a **0** in the other places.

ten-millions place
millions place
hundred-thousands place
ten-thousands place
thousands place
hundreds place
tens place
ones place

(4 3,5 7 0,0 0 0)

(2) Write the number that has a **6** in the ten-millions place, an **8** in the millions place, a **1** in the thousands place, a **2** in the tens place and a **0** in the other places.

(   )

(3) Write the number that you get from adding **2** ten-millions, **1** million, **7** hundred-thousands and **6** ten-thousands.

(   )

(4) Write the number that you get from adding **8** ten-millions, **5** millions and **1** ten-thousand.

(   )

(5) Write the number that you get from adding **9** millions and **3** hundred-thousands.

(   )

4 Write the appropriate number in each box.

5 points per question

(1) **70,000** is the number you get from adding [   ] ten-thousands.

(2) **170,000** is the number you get from adding [   ] ten-thousands.

(3) **170,000** is the number you get from adding [   ] thousands.

(4) **1,700,000** is the number you get from adding [   ] ten-thousands.

(5) **1,700,000** is the number you get from adding [   ] thousands.

(6) Three hundred fifty thousands is the same as the number you get from adding [   ] ten-thousands.

(7) The number you get from adding fifteen ten-thousands is [   ].

You're doing really well!

9

Large Numbers

1 Fill in the missing number in each box on the number lines below.

2 points per box

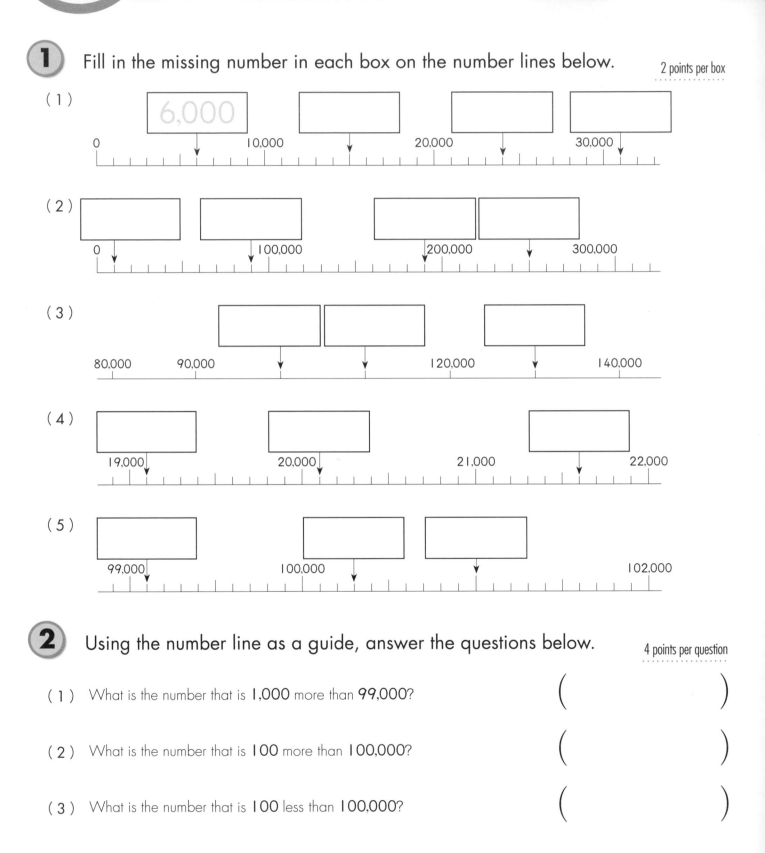

(1)

6,000 · · · · · · · · · · · ·

0 · · · 10,000 · · · 20,000 · · · 30,000

(2)

0 · · · 100,000 · · · 200,000 · · · 300,000

(3)

80,000 · · 90,000 · · · 120,000 · · · 140,000

(4)

19,000 · · · 20,000 · · · 21,000 · · · 22,000

(5)

99,000 · · · 100,000 · · · · · 102,000

2 Using the number line as a guide, answer the questions below.

4 points per question

(1) What is the number that is **1,000** more than **99,000**?   ( )

(2) What is the number that is **100** more than **100,000**?   ( )

(3) What is the number that is **100** less than **100,000**?   ( )

3 Write a ✓ under the larger number.

2 points per question

(1)

40,000	4,000
( )	( )

(2)

200,000	300,000
( )	( )

(3)

290,000	280,000
( )	( )

(4)

38,500	37,500
( )	( )

4 Rank the following groups of numbers from largest to smallest. Put a "1" next to the largest number, a "2" next to the next-largest number, and so on. 5 points per question

(1)

876,539	867,539	875,639
( )	( )	( )

(2)

99,999	100,200	100,120
( )	( )	( )

5 Compare the numbers below and write inequality signs (>, <) in the boxes.

4 points per question

(1) 500 [>] 300

(2) 8,000 [   ] 10,000

(3) 70,000 [   ] 60,000

(4) 35,000 [   ] 45,000

(5) 590,000 [   ] 570,000

(6) 2,440,000 [   ] 2,450,000

(7) 1,710,023 [   ] 1,701,023

(8) 101,010 [   ] 100,110

(9) 863,900 [   ] 864,000

`5 is larger than 4´ is written as `5 > 4,´ and `8 is less than 10´ is written as `8 < 10.´

Do you have your `less than´ and `greater than´ signs down? Good!

Large Numbers

6

Date / /

Name

1 Write the appropriate number in each box.

4 points per question

(1)   10 times 10 is ☐ .

(2)   10 times 100 is ☐ .

(3)   10 times 1,000 is ☐ .

(4)   10 times 110 is ☐ .

(5)   10 times 1,110 is ☐ .

(6)   10 times 52 is ☐ .

(7)   10 times 10 times 52 is ☐ .

2   Write the ten-times and hundred-times multiples of each number in the table.

8 points per question

(1)

Base number	38	47	60	500
ten times	380			
hundred times	3,800			

(2)

Base number	945	106	120	790
ten times				
hundred times				

3   Write the appropriate number in each box.

4 points per question

(1)   5,200 divided by 10 is ⬚ .

(2)   520 divided by 10 is ⬚ .

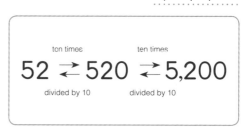

4   Divide each number below by 10.

6 points per question

(1)   70   (2)   80   (3)   400   (4)   3,000

(   )   (   )   (   )   (   )

(5)   190   (6)   610   (7)   9,200   (8)   9,020

(   )   (   )   (   )   (   )

Times ten or divided by ten—you can handle it!

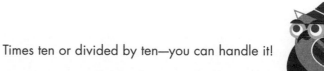

7 Fractions

Level ★★

Date   /   /

Name

Score   /100

1 When you divide a whole into two equal parts, each of those parts is called one half, and it is written $\frac{1}{2}$. Color in one half of the objects like the sample below.

5 points per question

⟨Example⟩

(1)

(2)

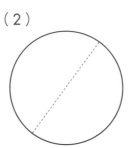

(3)

(4)

(5)

2 What fraction of each whole piece is shaded below?

5 points per question

(1)

$\left(\dfrac{1}{2} \right)$

(2)

$\left(\dfrac{1}{3} \right)$

(3)

$\left(\underline{} \right)$

(4)

$\left(\underline{} \right)$

(5)

$\left(\underline{} \right)$

(6)

$\left(\underline{} \right)$

(7)

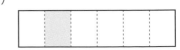

$\left(\underline{} \right)$

14   © Kumon Publishing Co., Ltd.

3 Each of these pieces of tape is one foot long and has been divided into equal parts. What fraction of the whole piece is shaded in each figure below?

5 points per question

(1)

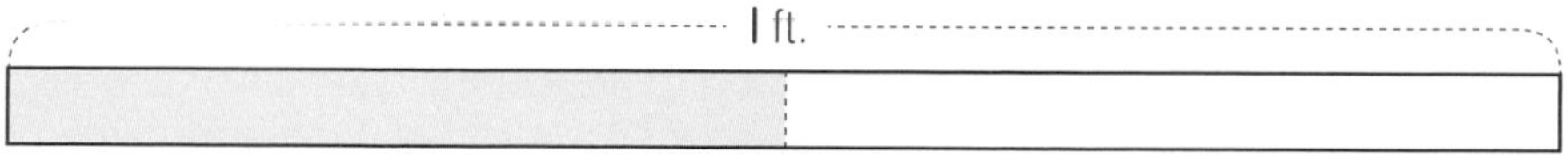

I ft.

$\left(\dfrac{1}{2}\right)$ft.

(2)

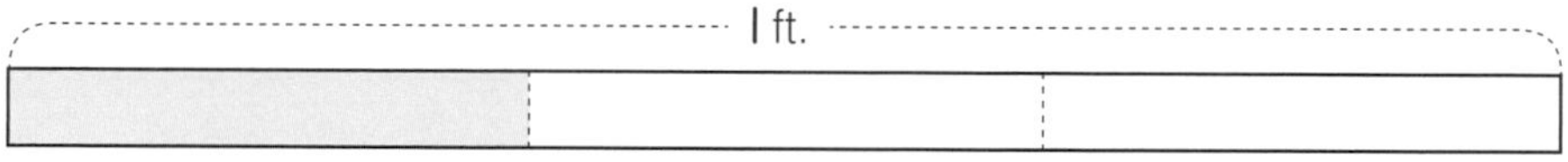

I ft.

$\left(\dfrac{}{}\right)$ft.

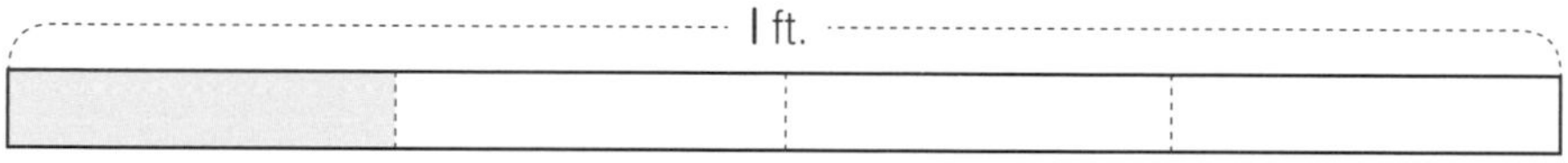

(3)

I ft.

$\left(\dfrac{}{}\right)$ft.

(4)

I ft.

$\left(\dfrac{}{}\right)$ft.

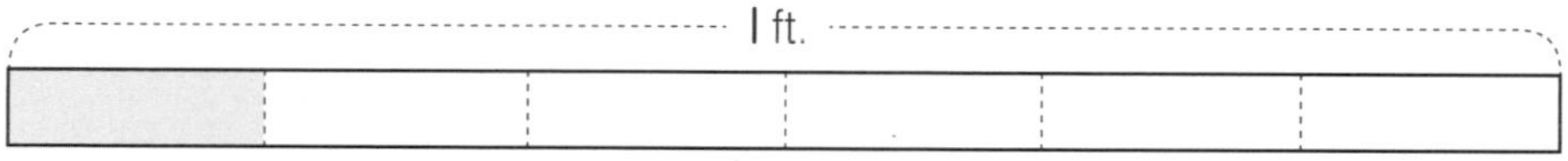

4 Each of these pieces of tape is one meter long and has been divided into equal parts. What fraction of the whole piece is shaded in each figure below?

5 points per question

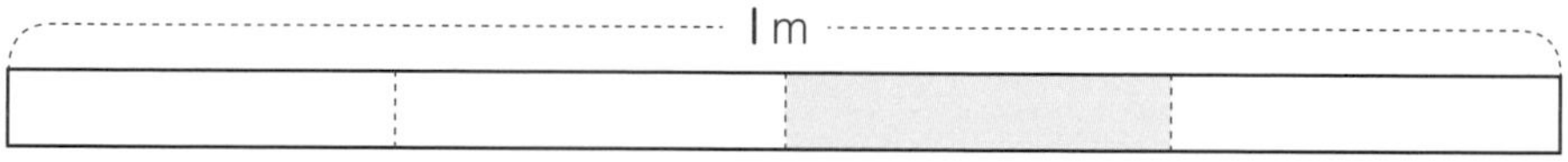

(1)

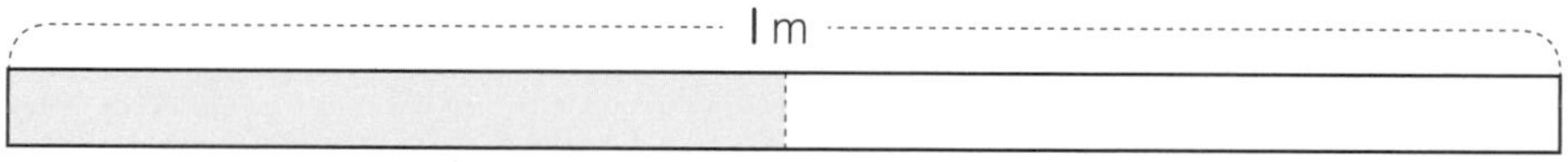

I m

$\left(\dfrac{1}{2}\right)$m

(2)

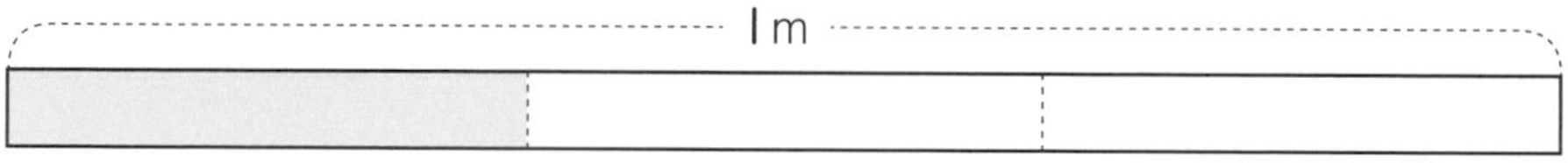

I m

$\left(\dfrac{}{}\right)$m

(3)

I m

$\left(\dfrac{}{}\right)$m

(4)

I m

$\left(\dfrac{}{}\right)$m

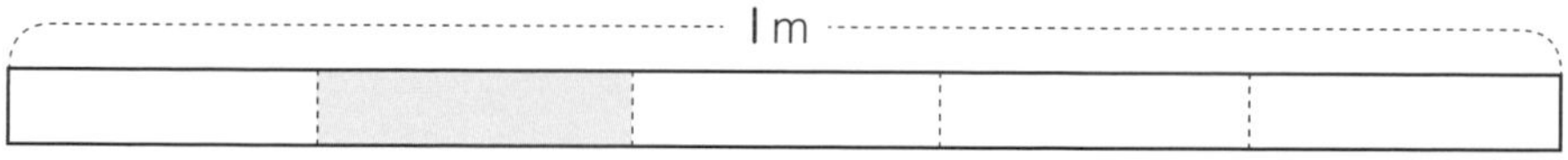

A *fraction* is a part of a whole. You got it.

8 Fractions

Date / /　　Name

Level ★★　　Score /100

Don't forget!

If a one-meter piece of tape is divided into 5 equal parts, and two pieces are kept, those pieces are $\frac{2}{5}$ of the original piece of tape.

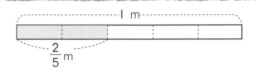

Numbers like $\frac{1}{2}$ and $\frac{2}{5}$ are called fractions.

In a fraction, the top number is called the numerator, and the bottom number is called the denominator.

$\frac{2}{5}$ ········ the numerator
········ the denominator

1　Write each fraction.

5 points per question

(1) A length of 1 is divided into 5 equal parts, and 3 parts are kept.

$\left(\ \frac{3}{5}\ \right)$

(2) A length of 1 is divided into 5 equal parts, and 4 parts are kept.

$\left(\ \frac{4}{5}\ \right)$

(3) A length of 1 is divided into 6 equal parts, and 5 parts are kept.

$\left(\ \frac{\ \ }{6}\ \right)$

(4) A length of 1 is divided into 7 equal parts, and 2 parts are kept.

$\left(\ \ \right)$

(5) A length of 1 is divided into 7 equal parts, and 4 parts are kept.

$\left(\ \ \right)$

(6) This fraction has a denominator of 6 and a numerator of 3.

$\left(\ \ \right)$

(7) This fraction has a denominator of 8 and a numerator of 5.

$\left(\ \ \right)$

(8) This fraction has a denominator of 9 and a numerator of 2.

$\left(\ \ \right)$

2 Write the number of shaded parts in each picture as a fraction.

5 points per question

(1)

$\left(\dfrac{\quad}{4} \right)$

(2)

$\left(\dfrac{\quad}{4} \right)$

(3)

$\left(\phantom{\dfrac{x}{x}} \right)$

(4)

$\left(\phantom{\dfrac{x}{x}} \right)$

(5)

$\left(\phantom{\dfrac{x}{x}} \right)$

3 Write the number of shaded parts of each foot-long ruler as a fraction.

5 points per question

(1)

— I ft. —

$\left(\dfrac{\quad}{6} \right)$ ft.

(2)

— I ft. —

$\left(\dfrac{\quad}{6} \right)$ ft.

(3)

— I ft. —

$\left(\phantom{\dfrac{x}{x}} \right)$ ft.

(4)

— I ft. —

$\left(\phantom{\dfrac{x}{x}} \right)$ ft.

(5)

— I ft. —

$\left(\phantom{\dfrac{x}{x}} \right)$ ft.

(6)

— I ft. —

$\left(\phantom{\dfrac{x}{x}} \right)$ ft.

(7)

— I ft. —

$\left(\phantom{\dfrac{x}{x}} \right)$ ft.

This is tough, but you're doing great!

Don't forget!

The fractions of an inch are represented below.

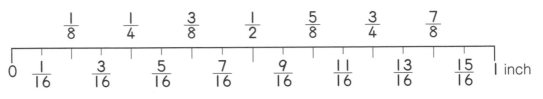

1 What fraction of an inch is each letter pointing to?

4 points per letter

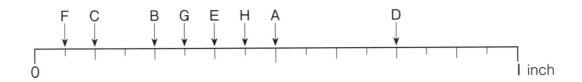

A ( )   B ( )   C ( )   D ( )

E ( )   F ( )   G ( )   H ( )

Don't forget!

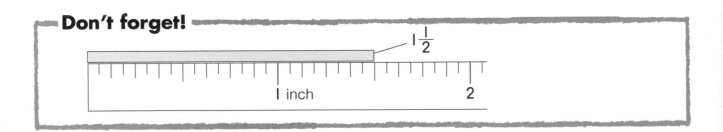

2 How many inches long is each line?

4 points per question

(1)

(1 in.)

(2)

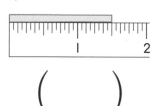

(   )

(3)

(   )

(4)

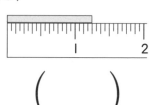

(   )

(5)

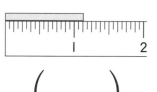

(   )

(6)

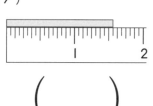

(   )

(7)

(   )

(8)

(   )

3 How many inches long is each line?

6 points per question

(1)

(   )

(2)

(   )

(3)

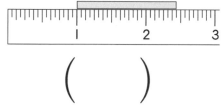

(   )

(4)

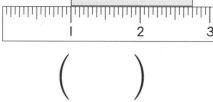

(   )

(5)

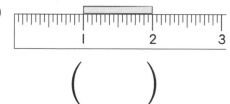

(   )

(6)

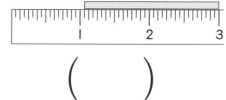

(   )

Each of those little lines on the ruler is a fraction of an inch! You got it.

Date / /

Name

Level ⭐⭐

Score /100

Don't forget!

There are 5,280 feet in one mile, which is written "1 mi." (2 mi.=10,560 ft., 3 mi.=15,840 ft.,······)

1 Write the numbers in each box.

3 points per question

(1) 1 mi. = ☐ ft.

(2) 1 mi. 5 ft. = ☐ ft.

(3) 1 mi. 50 ft. = ☐ ft.

(4) 1 mi. 500 ft. = ☐ ft.

(5) 1 mi. 540 ft. = ☐ ft.

(6) 2 mi. = ☐ ft.

(7) 2 mi. 70 ft. = ☐ ft.

(8) 2 mi. 80 ft. = ☐ ft.

(9) 2 mi. 300 ft. = ☐ ft.

(10) 3 mi. 20 ft. = ☐ ft.

(11) 5,280 ft. = ☐ mi.

(12) 5,290 ft. = ☐ mi. ☐ ft.

(13) 5,500 ft. = ☐ mi. ☐ ft.

(14) 6,000 ft. = ☐ mi. ☐ ft.

(15) 7,000 ft. = ☐ mi. ☐ ft.

(16) 9,000 ft. = ☐ mi. ☐ ft.

(17) 10,560 ft. = ☐ mi.

(18) 11,000 ft. = ☐ mi. ☐ ft.

(19) 13,000 ft. = ☐ mi. ☐ ft.

(20) 15,840 ft. = ☐ mi.

2 Write a ✓ under the longer distance.

(1)　　900 ft.　　　　800 ft.

(　　)　(　　)

(2)　　720 ft.　　　　702 ft.

(　　)　(　　)

(3)　　 I mi.　　　　 5,270 ft.

(　　)　(　　)

(4)　 I mi. 200 ft.　　 I mi. 300 ft.

(　　)　(　　)

(5)　 I mi. 80 ft.　　 I mi. I00 ft.

(　　)　(　　)

(6)　 I mi. 850 ft.　　 I mi. 950 ft.

(　　)　(　　)

(7)　 I mi. 200 ft.　　 6,000 ft.

(　　)　(　　)

(8)　 I mi. 240 ft.　　 6,500 ft.

(　　)　(　　)

(9)　 5,200 ft.　　 I mi. 40 ft.

(　　)　(　　)

(10)　 5,300 ft.　　 I mi. I0 ft.

(　　)　(　　)

3 Rank the following groups of distances from longest to shortest.

(1)　 I mi.　　 5,400 ft.　　 I mi. 20 ft.　　 5,270 ft.

(　　)　(　　)　(　　)　(　　)

(2)　 I0,500 ft.　　 2 mi.　　 I mi. 500 ft.　　 I mi. 5,000 ft.

(　　)　(　　)　(　　)　(　　)

5,280 feet seems like a lot more than a mile, doesn't it? But it is not!

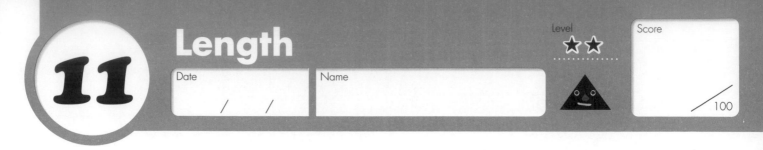

Length

Date   /   /

Name

Level ★★

Score   /100

1 10mm=1cm. How many millimeters is it from the left side of the ruler to each box?

3 points per question

(1)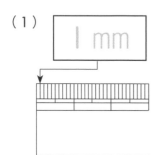

(2)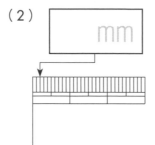

(3)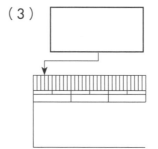

(4)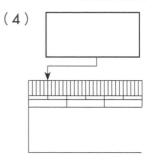

(5)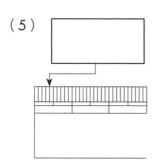

(6)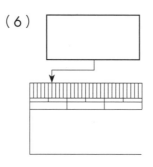

(7)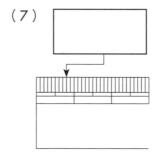

(8)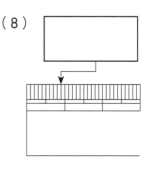

2 How many millimeters is it from the left side of the ruler to each box?

3 points per question

(1)

(2)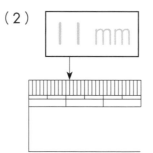

(3)

(4)

(5)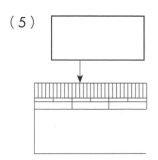

(6)

(7)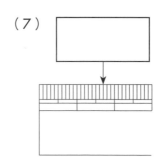

(8)

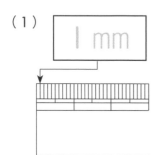

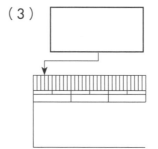
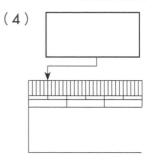

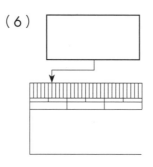

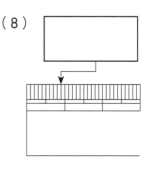
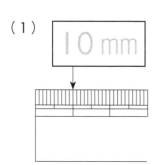
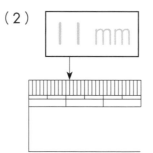
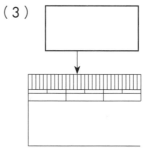
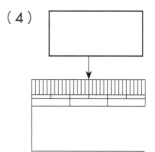
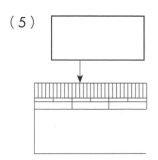
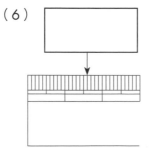
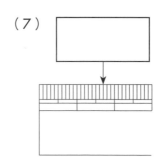
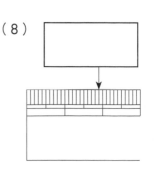

3 1 cm = 10 mm. How many millimeters is it from the left side of the ruler to each box?

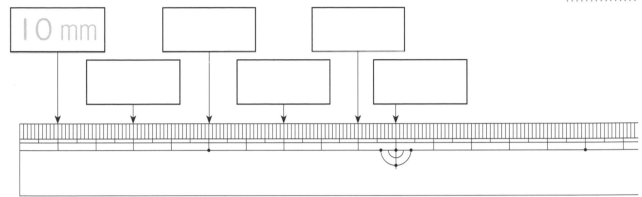

4 How many millimeters long is each line below?

(1)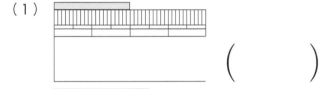

(   )

(2)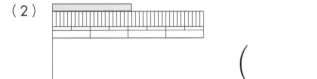

(   )

(3)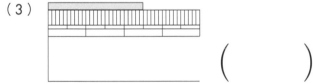

(   )

(4)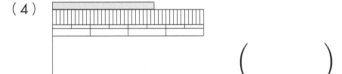

(   )

(5)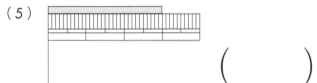

(   )

(6)

(   )

5 How many millimeters thick is each book below?

(1)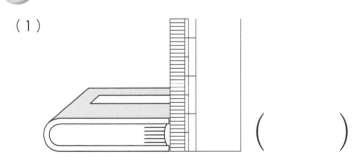

(   )

(2)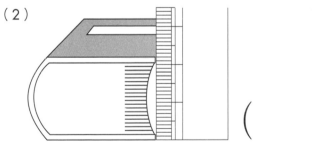

(   )

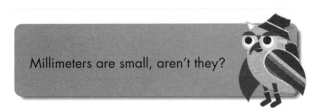

Millimeters are small, aren't they?

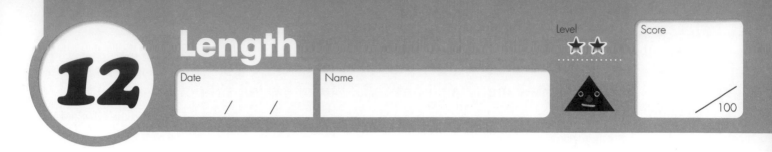

Length

12

Date   /   /

Name

Level ☆☆

Score   /100

---

1   How far is each box from the left side of the ruler?

5 points per box

(1)

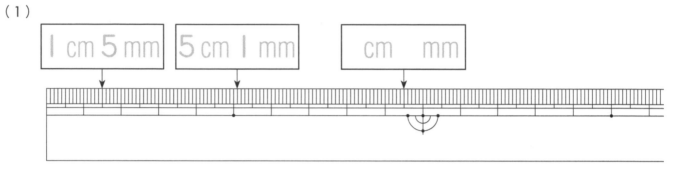

| 1 cm 5 mm | 5 cm 1 mm | cm   mm |

(2)

---

2   How long is each line?

10 points per question

(1)

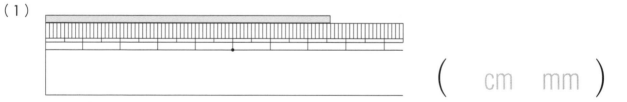

(cm   mm   )

(2)

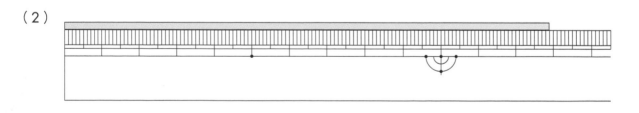

(cm   mm   )

3 How long is each piece of tape?

(1)

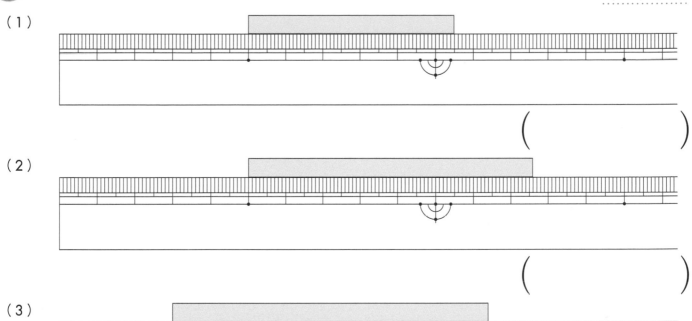

(   )

(2)

(   )

(3)

(   )

4 Along each dashed line, draw a line that fits the measurement given below.

5 points per question

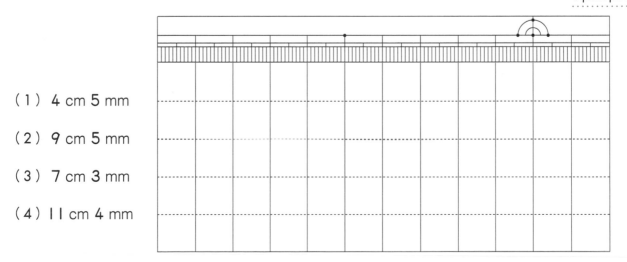

(1) 4 cm 5 mm

(2) 9 cm 5 mm

(3) 7 cm 3 mm

(4) 11 cm 4 mm

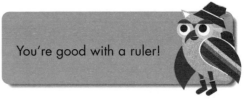

You're good with a ruler!

Date / /

Name

Level ⭐⭐

Score /100

Don't forget!

There are 1,000 meters (m) in 1 kilometer, which is written "1 km".

---

1 Convert the measurements below.

3 points per question

(1) 1 km = [1,000] m

(2) 1 km 5 m = [ ] m

(3) 1 km 50 m = [ ] m

(4) 1 km 500 m = [ ] m

(5) 1 km 550 m = [ ] m

(6) 2 km = [ ] m

(7) 2 km 80 m = [ ] m

(8) 2 km 400 m = [ ] m

(9) 2 km 650 m = [ ] m

(10) 3 km 180 m = [ ] m

(11) 1,000 m = [1] km

(12) 1,400 m = [ ] km [ ] m

(13) 1,060 m = [ ] km [ ] m

(14) 1,850 m = [ ] km [ ] m

(15) 2,300 m = [ ] km [ ] m

(16) 2,050 m = [ ] km [ ] m

(17) 2,560 m = [ ] km [ ] m

(18) 3,000 m = [ ] km

(19) 3,200 m = [ ] km [ ] m

(20) 3,650 m = [ ] km [ ] m

2 Write a ✓ under the longer distance.

3 points per question

(1)
700 m	800 m
( )	( )

(2)
650 m	605 m
( )	( )

(3)
1 km	950 m
( )	( )

(4)
1 km 300 m	1 km 400 m
( )	( )

(5)
1 km 70 m	1 km 100 m
( )	( )

(6)
1 km 550 m	1 km 650 m
( )	( )

(7)
1 km 200 m	1,100 m
( )	( )

(8)
1 km 350 m	1,380 m
( )	( )

(9)
1,600 m	1 km 80 m
( )	( )

(10)
2,050 m	2 km 500 m
( )	( )

3 Rank the distances below from longest to shortest.

5 points per question

(1)
1 km	1,100 m	1 km 10 m	990 m
( )	( )	( )	( )

(2)
1,880 m	2 km	1 km 790 m	2,050 m
( )	( )	( )	( )

Do you run the 100 m in track?
I bet you're fast.

27

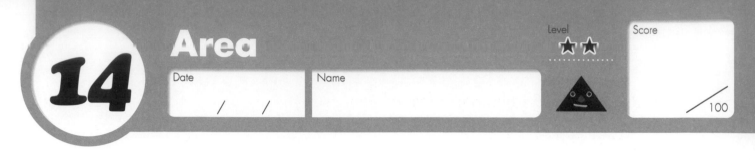

14 Area

Level ☆☆ Score /100

Date   /   /

Name

---

1 The scale of the grid on the right is 1 inch.

10 points per question

(1) How many squares with 1-inch sides does each figure contain?

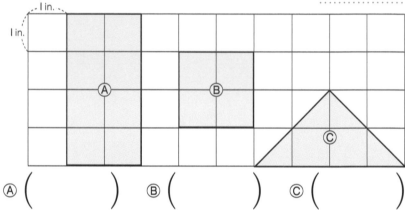

Ⓐ (   ) Ⓑ (   ) Ⓒ (   )

(2) Which figure is the largest?

(   )

Don't forget!

A square with sides of 1 inch has an area of one square inch, which is written 1 in.²

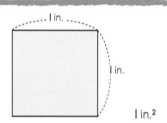

1 in.²

---

2 The figures below are made by squares that have 1-inch sides. How many square inches is the area of each figure?

5 points per question

(1)

(1 in.²)

(2)

(   )

(3)

(   )

(4)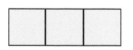

(   )

(5)

(   )

(6)

(   )

---

3 The scale of the grid on the right is 1 cm.

(1) How many squares with 1-cm sides does each figure contain?

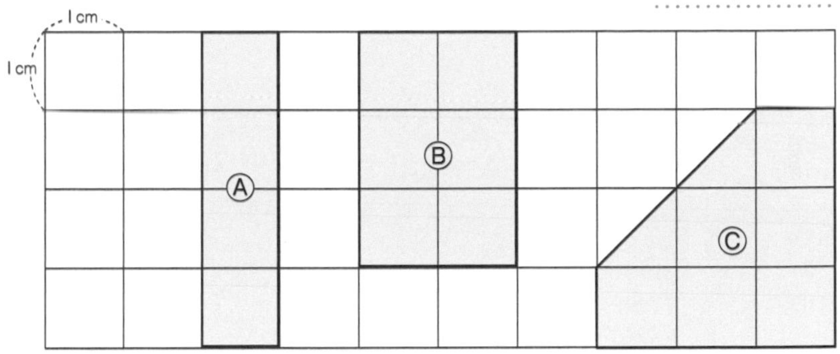

Ⓐ (   )   Ⓑ (   )   Ⓒ (   )

(2) Which figure is the largest?

(   )

Don't forget!

A square with sides of 1 cm has an area of one square centimeter, which is written 1 cm².

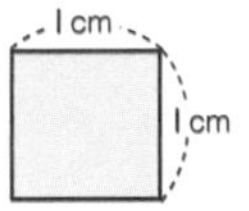

1 cm²

---

4 The figures below are made by squares that have 1-cm sides. How many square centimeters is the area of each figure?

(1)

(1 cm²)

(2)

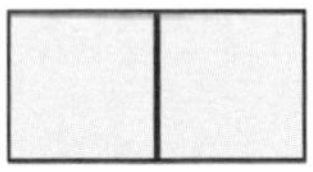

(2 cm²)

(3)

(   )

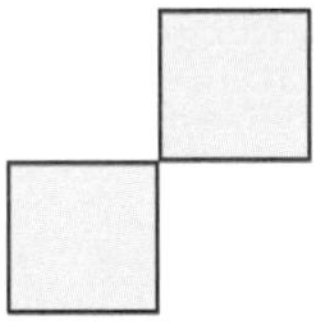

(4)

(   )

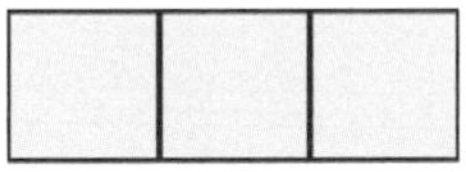

(5)

(   )

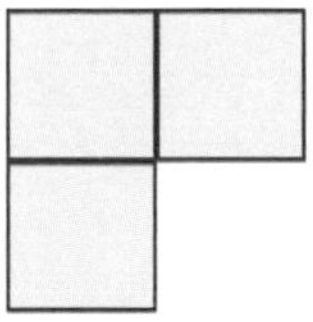

(6)

(   )

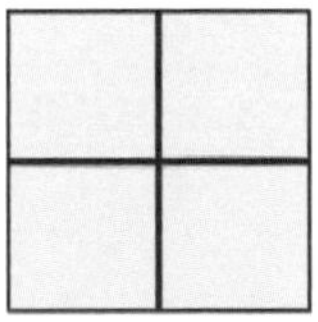

'Area' is another way of describing the flat space a figure covers. Good job!

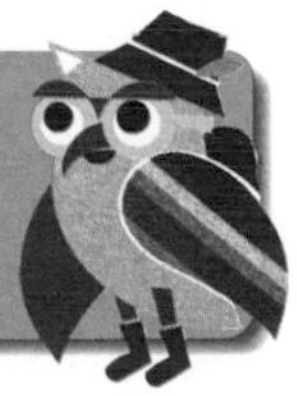

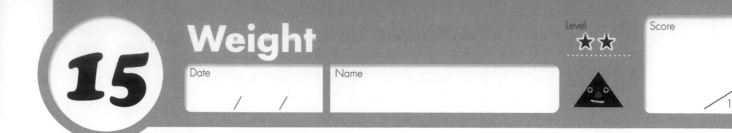

Weight

15

Level ★★

Score

Date / /

Name

/100

1 What is the weight on each scale below?

5 points per question

(1)
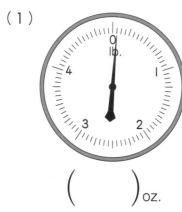

(   ) oz.

(2)
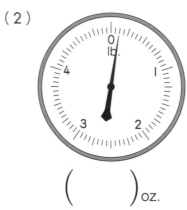

(   ) oz.

(3)
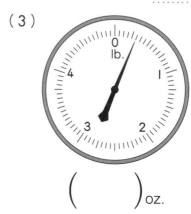

(   ) oz.

(4)
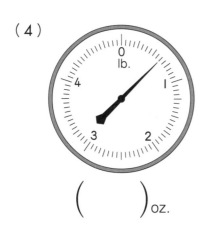

(   ) oz.

(5)
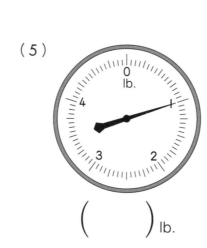

(   ) lb.

(6)
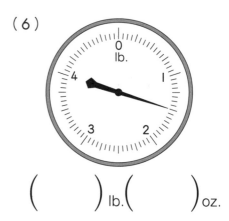

(   ) lb. (   ) oz.

(7)
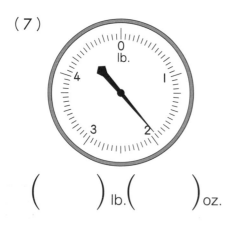

(   ) lb. (   ) oz.

(8)
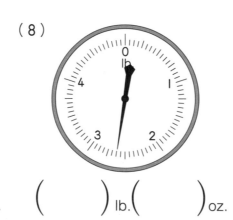

(   ) lb. (   ) oz.

Don't forget!

16 ounces = 1 pound
(oz.)   (lb.)

2 Convert the measurements below.

3 points per question

(1) 1 lb. = ☐ oz.

(2) 1 lb. 1 oz. = ☐ oz.

(3) 1 lb. 3 oz. = ☐ oz.

(4) 1 lb. 5 oz. = ☐ oz.

(5) 1 lb. 10 oz. = ☐ oz.

(6) 1 lb. 15 oz. = ☐ oz.

(7) 2 lb. = ☐ oz.

(8) 2 lb. 2 oz. = ☐ oz.

(9) 2 lb. 5 oz. = ☐ oz.

(10) 3 lb. 4 oz. = ☐ oz.

(11) 16 oz. = ☐ lb.

(12) 17 oz. = ☐ lb. ☐ oz.

(13) 18 oz. = ☐ lb. ☐ oz.

(14) 20 oz. = ☐ lb. ☐ oz.

(15) 26 oz. = ☐ lb. ☐ oz.

(16) 27 oz. = ☐ lb. ☐ oz.

(17) 30 oz. = ☐ lb. ☐ oz.

(18) 40 oz. = ☐ lb. ☐ oz.

(19) 42 oz. = ☐ lb. ☐ oz.

(20) 45 oz. = ☐ lb. ☐ oz.

I wonder how much your shoes weigh.
Good job!

31

16

Weight

Date ___/___/___

Name

Level ★★

Score
____/100

1 What is the weight on each scale below?

5 points per question

(1)

(200 g)

(2)
(___ g)

(3)

(___ g)

(4)
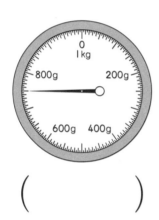
( )

(5)
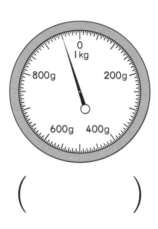
( )

(6)
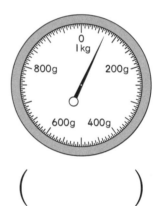
( )

(7)
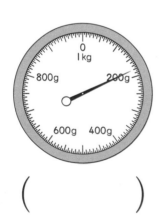
( )

(8)
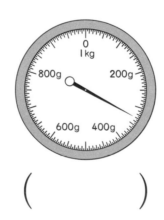
( )

(9)
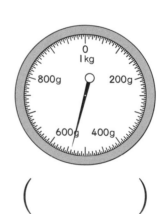
( )

2 What is the weight on each scale below?

5 points per question

(1)

(10 g)

(2)

( )

(3)

( )

(4)
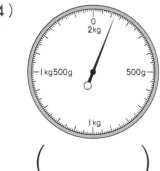
( )

(5)
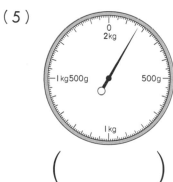
( )

(6)

( )

(7)
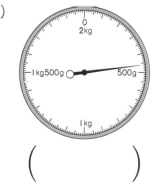
( )

(8)

( )

(9)
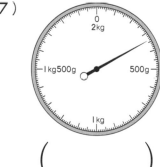
( )

(10)
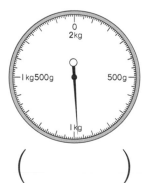
( )

(11)
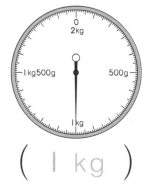
(1 kg)

You're a pro!

33

17 Weight

Level ☆☆

Date / /

Name

Score /100

Don't forget!

There are 1,000 grams (g) in 1 kilogram, which is written "1 kg".

1 Convert the measurements below.

3 points per question

(1) 1 kg = [1,000] g

(2) 1 kg 50 g = [ ] g

(3) 1 kg 500 g = [ ] g

(4) 2 kg = [ ] g

(5) 2 kg 10 g = [ ] g

(6) 2 kg 100 g = [ ] g

(7) 2 kg 800 g = [ ] g

(8) 3 kg = [ ] g

(9) 3 kg 40 g = [ ] g

(10) 3 kg 400 g = [ ] g

(11) 1,000 g = [1] kg

(12) 1,200 g = [1] kg [200] g

(13) 1,060 g = [ ] kg [ ] g

(14) 2,100 g = [ ] kg [ ] g

(15) 2,080 g = [ ] kg [ ] g

(16) 2,500 g = [ ] kg [ ] g

(17) 3,000 g = [ ] kg

(18) 3,600 g = [ ] kg [ ] g

(19) 4,000 g = [ ] kg

(20) 4,080 g = [ ] kg [ ] g

2 Write a ✓ under the heavier weight.

3 points per question

(1)
800 g	600 g
( )	( )

(2)
550 g	505 g
( )	( )

(3)
I kg	900 g
( )	( )

(4)
I kg 200 g	I kg 300 g
( )	( )

(5)
I kg 100 g	I kg 90 g
( )	( )

(6)
I kg	1,100 g
( )	( )

(7)
I kg 500 g	1,400 g
( )	( )

(8)
2 kg 50 g	2,500 g
( )	( )

(9)
2,080 g	2 kg 10 g
( )	( )

(10)
4,600 g	4 kg 90 g
( )	( )

3 Rank the weights below from heaviest to lightest.

5 points per question

(1)
I kg	1,010 g	I kg 100 g	990 g
( )	( )	( )	( )

(2)
3 kg 700 g	4 kg	3,800 g	3,090 g
( )	( )	( )	( )

Does your scale at home have kilograms on it?
Or pounds?

18
Weight

Date   /   /

Name

Level ★★

Score

/100

1 What is the weight on each scale below?

8 points per question

(1)

(1 kg 200 g)

(2)

(kg   g)

(3)

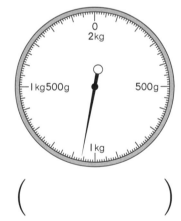

(   )

(4)

(   )

(5)

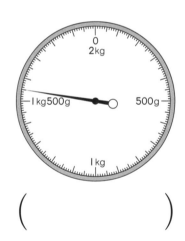

(   )

(6)

(   )

(7)

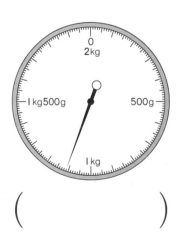

(   )

(8)

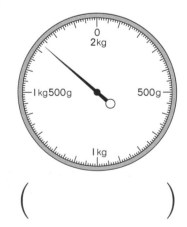

(   )

2 What is the weight on each scale below?

6 points per question

(1)

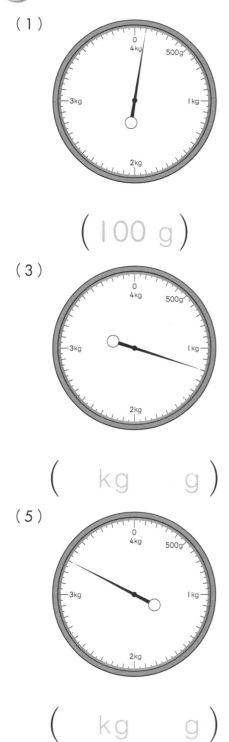

(100 g)

(2)

(g)

(3)

(kg   g)

(4)

(kg   g)

(5)

(kg   g)

(6)

(kg   g)

What kind of scale do they use at the grocery store? Take a look next time!

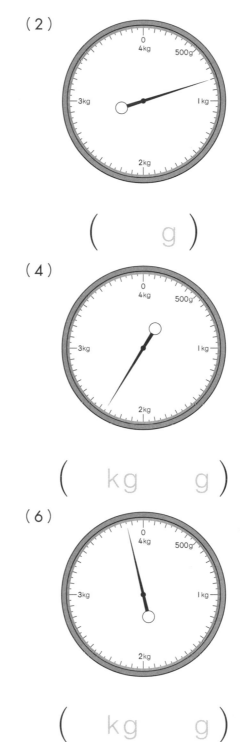

Weight

Don't forget!

1,000 mg = 1 g

1 What is the weight on each scale below?

6 points per question

(1)

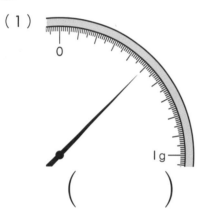

0

1 g

(   )

(2)

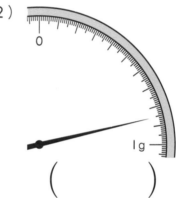

0

1 g

(   )

2 Convert the measurements below.

5 points per question

(1) 1 g = | 1,000 | mg

(2) 2 g = | | mg

(3) 4 g = | | mg

(4) 6 g = | | mg

(5) 10 g = | | mg

(6) 1,000 mg = | 1 | g

(7) 3,000 mg = | | g

(8) 5,000 mg = | | g

(9) 7,000 mg = | | g

(10) 9,000 mg = | | g

3 Write a ✓ under the heavier weight.

3 points per question

(1)
700 mg	800 mg
( )	( )

(2)
350 mg	305 mg
( )	( )

(3)
950 mg	1 g
( )	( )

(4)
2,010 mg	2 g
( )	( )

(5)
1,100 mg	1,101 mg
( )	( )

(6)
3,010 mg	3,001 mg
( )	( )

4 Rank the following groups from heaviest to lightest.

5 points per question

(1)
1 g	1,001 mg	995 mg	1,100 mg
( )	( )	( )	( )

(2)
1 kg	1 mg	1 g	10,000 mg
( )	( )	( )	( )

(3)
2,500 mg	3 g	2,010 mg	2 g
( )	( )	( )	( )

(4)
10 mg	1 mg	1 g	100 mg
( )	( )	( )	( )

Weighing is fun, isn't it?

Capacity

Date   /   /

Name

Level ★★

Score   /100

1 Look at the water in each of the cups below. Then rank them from most water to least water.

10 points per question

(1) Ⓐ 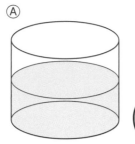 (   )   Ⓑ 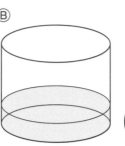 (   )   Ⓒ 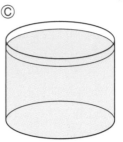 (   )

(2) Ⓐ 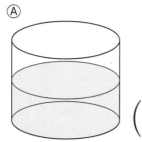 (   )   Ⓑ  (   )   Ⓒ  (   )

(3) Ⓐ  (   )   Ⓑ  (   )   Ⓒ  (   )

(4) Ⓐ  (   )   Ⓑ 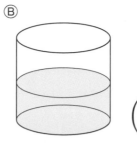 (   )   Ⓒ 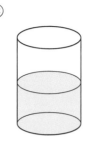 (   )

© Kumon Publishing Co., Ltd.

2 Ted and Bob filled some cups with their canteens. Write a ✓ under the canteen that had more water.

20 points

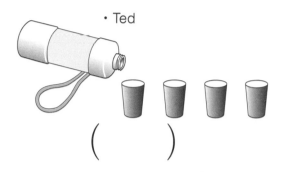 · Ted

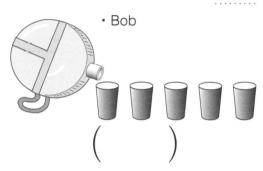 · Bob

(   )

(   )

3 We filled some cups using a kettle of water and a bottle of water. Which one holds more cups of water? How many more cups does it hold?

20 points

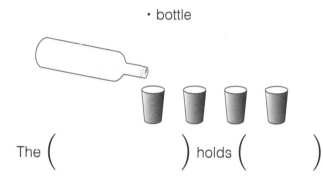 · bottle

 · kettle

The (   ) holds (   ) more cups.

4 Rank the following groups of cups from most to least.

20 points

  (   )

  (   )

  (   )

The amount of liquid a container can hold is called *capacity*.

21 Capacity

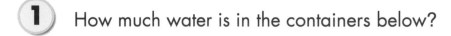

Date / /

Name

Level ★★

Score /100

---

1 How much water is in the containers below?

5 points per question

(1)

I pint

(I pt.)

(2)

I pint I pint

( )

(3)

I pint I pint I pint

( )

(4)

I pint I pint I pint

I pint I pint

( )

---

2 How much water is in the containers below?

5 points per question

(1)

I quart

(qt.)

(2)

I quart I quart

( )

(3)

I quart I quart

I quart I quart

( )

(4)

I quart I quart I quart

I quart I quart I quart

( )

Don't forget!

2 pints equal 1 quart, which is written "1 qt."

2 pt. = 1 qt.

3 Convert the measurements below.

3 points per question

(1) 1 qt. = ☐ pt.

(2) 1 qt. 1 pt. = ☐ pt.

(3) 2 qt. = ☐ pt.

(4) 3 qt. = ☐ pt.

(5) 3 qt. 1 pt. = ☐ pt.

(6) 4 qt. 1 pt. = ☐ pt.

(7) 5 qt. = ☐ pt.

(8) 6 qt. = ☐ pt.

(9) 6 qt. 1 pt. = ☐ pt.

(10) 7 qt. 1 pt. = ☐ pt.

(11) 2 pt. = ☐ qt.

(12) 4 pt. = ☐ qt.

(13) 5 pt. = ☐ qt. ☐ pt.

(14) 8 pt. = ☐ qt.

(15) 11 pt. = ☐ qt. ☐ pt.

(16) 14 pt. = ☐ qt.

(17) 15 pt. = ☐ qt. ☐ pt.

(18) 16 pt. = ☐ qt.

(19) 17 pt. = ☐ qt. ☐ pt.

(20) 20 pt. = ☐ qt.

A big glass is about a pint.
Ask your parents to show you a pint!

Capacity

1 How much water is in the containers below?

5 points per question

(1)

( )

(2)

( )

(3)

( )

(4)

( )

> **Don't forget!**
> 4 quarts equal 1 gallon, which is written "1 gal."
> 4 qt. = 1 gal.

2 Convert the measurements below.

5 points per question

(1) 1 gal. = ☐ qt.

(2) 1 gal. 1 qt. = ☐ qt.

(3) 1 gal. 3 qt. = ☐ qt.

(4) 2 gal. = ☐ qt.

(5) 4 qt. = ☐ gal.

(6) 6 qt. = ☐ gal. ☐ qt.

(7) 9 qt. = ☐ gal. ☐ qt.

(8) 10 qt. = ☐ gal. ☐ qt.

3 How much water is in the containers below? Write each measurement two or three different ways.

5 points per question

(1)

_____gal. _____qt.

_____qt.

(2)

_____gal._____qt.

_____qt.

(3)

_____gal.

_____qt.

(4)

_____gal._____qt.

_____qt.

(5)

_____qt._____pt.

_____pt.

(6)

_____qt.

_____pt.

(7)

_____qt.

_____pt.

(8)

I gallon

_____gal.

_____qt.

_____pt.

Don't forget!

I gal. = 4 qt.   I gal. = 8 pt.
I qt. = 2 pt.

You're doing really well!

45

23 Capacity

Level ☆☆

Date / /

Name

Score /100

1 How much water is in the containers below?

5 points per question

(1)

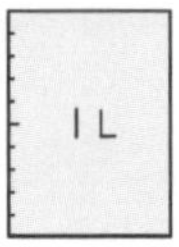

(1 L)

(2)

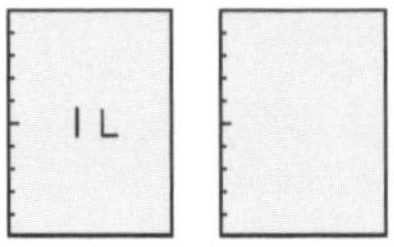

(L)

(3)

( )

(4)

( )

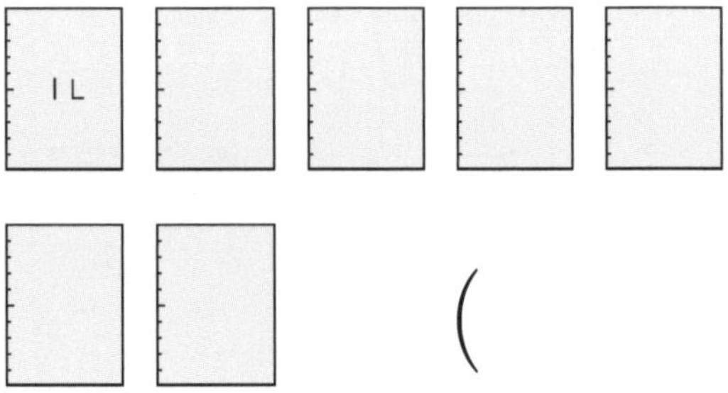

Don't forget!

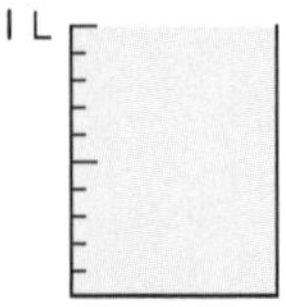

1 L = 1,000 mL
1 line on the container = 100 mL

2 Convert the measurements below.

4 points per question

(1) 1 L = [1,000] mL

(2) 2 L = [ ] mL

(3) 4 L = [ ] mL

(4) 8 L = [ ] mL

(5) 1,000 mL = [1] L

(6) 3,000 mL = [ ] L

(7) 9,000 mL = [ ] L

(8) 11,000 mL = [ ] L

3 Remember that 1 L = 1,000 mL. How much water is in the containers below?

4 points per question

(1) (100 mL)

(2) (mL)

(3) (mL)

(4) (   )

(5) (   )

(6) (   )

(7) (   )

4 How much water is in the containers below?

5 points per question

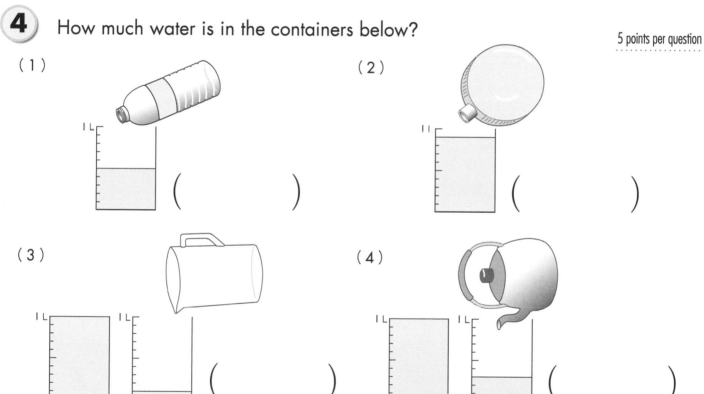

(1) (   )

(2) (   )

(3) (   )

(4) (   )

Do you have any drinks in your fridge that say liters or milliliters on them? Go check!

Telling Time

Date   /   /

Name

Level ★★

Score

/100

1 What time is it? Write the time under each clock below.

5 points per question

(1)

(8:00)

(2)

(8:01)

(3)

(8:   )

(4)

(   )

(5)

(   )

(6)

(   )

2 Write the time under each clock below.

5 points per question

(1)

(   )

(2)

(   )

(3)

(   )

(4)

(   )

3 Write the time under each clock below.

5 points per question

(1)

(8:05)

(2)

(8:06)

(3)

(8:07)

(4)

(　　　)

(5)

(　　　)

(6)

(　　　)

4 Write the time under each clock below.

5 points per question

(1)

(　　　)

(2)

(　　　)

(3)

(　　　)

(4)

(　　　)

Do you have a watch? Go find it!

Telling Time

Date / /
Name

Level

Score
/100

1 Write the time under each clock below.

5 points per question

(1)

(2)

(3)

(4)

(8:05) (:05) (   ) (   )

2 Write the time under each clock below.

5 points per question

(1)

(2)

(3)

(   ) (   ) (   )

(4)

(5)

(6)

(   ) (   ) (   )

3 Write the time under each clock below.

5 points per question

(1)

(8 : 15)

(2)

(8 :)

(3)

(:)

(4)

( )

(5)

( )

(6)

( )

4 Write the time under each clock below.

5 points per question

(1)

( )

(2)

( )

(3)

( )

(4)

( )

You're getting good at this!

Telling Time

1 Write the time under each clock below.

4 points per question

(1)

(8:05)

(2)

( )

(3)

( )

(4)

( )

(5)

(8:25)

(6)

(8:30)

(7)

( )

(8)

( )

(9)

(8:45)

(10)

(8:50)

(11)

( )

(12)

( )

2 Write the time under each clock below.

4 points per question

(1)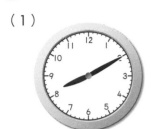

(2)

(3)

(4)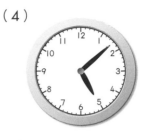

(8 : 10) (: 10) ( ) ( )

3 Write the time under each clock below.

6 points per question

(1)

( )

(2)

( )

(3)

( )

(4)

( )

(5)

( )

(6)

( )

What time is it now?

53

27 Telling Time

Date / /

Name

1 Write the time under each clock below.

5 points per question

(1)

(8:10)

(2)

(8:11)

(3)

(8:)

(4)

( )

(5)

( )

(6)

( )

2 Write the time under each clock below.

5 points per question

(1)

( )

(2)

( )

(3)

( )

(4)

( )

3 Write the time under each clock below.

5 points per question

(1)

(8:15)

(2)

(8:)

(3)

(:)

(4)

( )

(5)

( )

(6)

( )

4 Write the time under each clock below.

5 points per question

(1)

( )

(2)

( )

(3)

( )

(4)

( )

Now you've got it!

28 Telling Time

Level ☆☆

Date / /

Name

Score /100

1 Write the time under each clock below.

5 points per question

(1)

(9:17)

(2)

( )

(3)

( )

(4)

( )

2 Write the time under each clock below.

5 points per question

(1)

( )

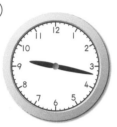

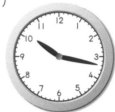

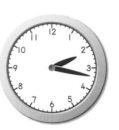

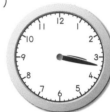

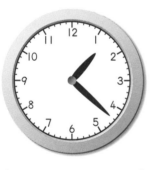

(2)

( )

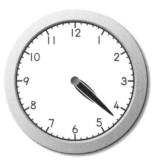

(3)

( )

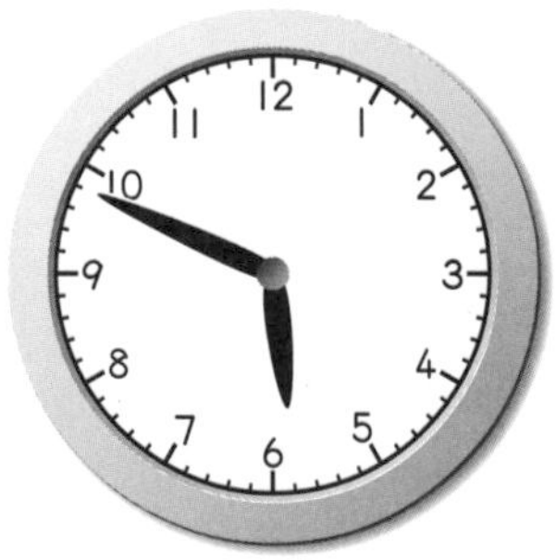

(4)

( )

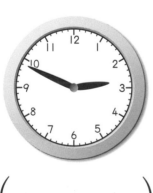

(5)

( )

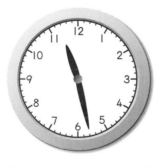

(6)

( )

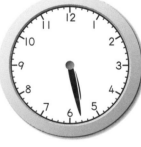

3 Write the time under each clock below.

5 points per question

(1) (2) (3) (4)

(3:47) (   ) (   ) (   )

4 Write the time under each clock below.

5 points per question

(1) (2) (3)

(   ) (   ) (   )

(4) (5) (6)

(   ) (   ) (   )

Ready for something a little different? Good!

29 Triangles & Quadrilaterals

Level ★★

Date / / Name

Score /100

Don't forget!

The circled angles in the triangles on the left are called **right angles**.

1 Circle the right angles in the triangles below.

20 points for completion

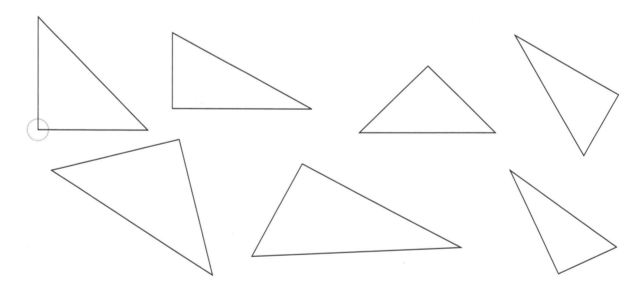

2 Take a look at the angles pictured here. Write a ✓ under the angle if it is a right angle, and × if it is not a right angle.

10 points per question

(1)

( )

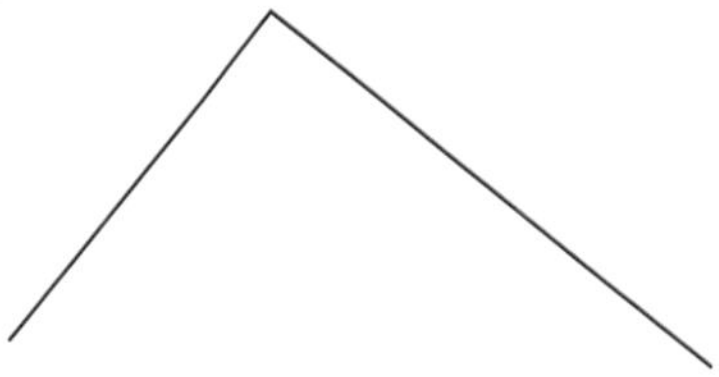

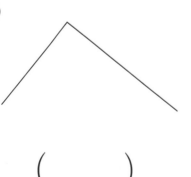

(2)

( )

(3)

( )

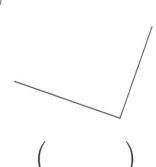

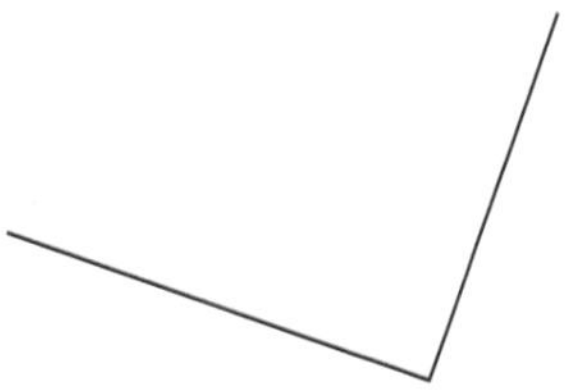

Don't forget!

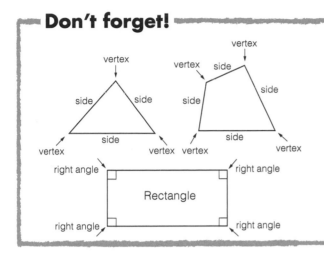

The straight lines that make up triangles and quadrilaterals are called **sides**. Each point where those lines come together is called a **vertex**. The plural of vertex is **vertices**.

A quadrilateral with four right angles is a **rectangle**. Each point where the lines come together is also called a vertex.

3 Look at the rectangle pictured here and answer the questions.

10 points per question

Rectangle

(1) How many vertices does the rectangle have? ( )

(2) How many right angles does the rectangle have? ( )

(3) How many pairs of sides with the same length does the rectangle have? ( )

4 Find the five rectangles in the shapes pictured here and write their letters below.

20 points for completion

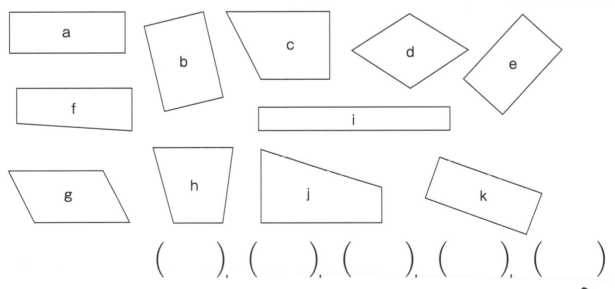

( ), ( ), ( ), ( ), ( )

Do you know what a rectangle is now? Good job!

Date / /

Name

Don't forget!

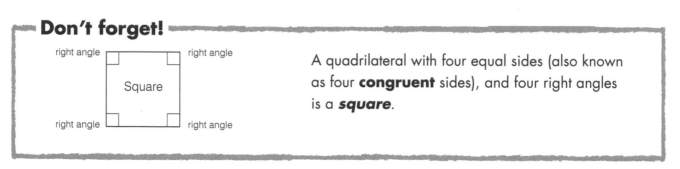

right angle   right angle

Square

right angle   right angle

A quadrilateral with four equal sides (also known as four **congruent** sides), and four right angles is a **square**.

1 Look at the square pictured here and answer the questions.

10 points per question

Square

(1) How many right angles does the square have?

(   )

(2) Are the four sides the same length?

(   )

2 Find the three squares in the shapes pictured here and write their letters below.

15 points for completion

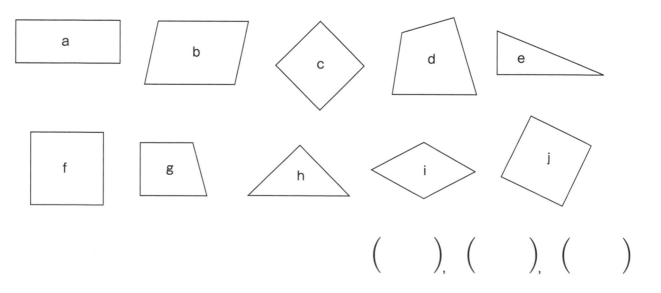

(   ) , (   ) , (   )

3 Sort the shapes pictured here into the categories below.

15 points per question

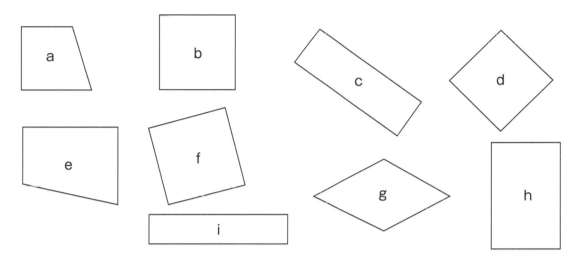

(1) Square ·· (   )

(2) Rectangle ··· (   )

(3) Not a square or a rectangle ····················· (   )

4 The rectangle below is folded to create shape A and then cut to create shape B.

10 points per question

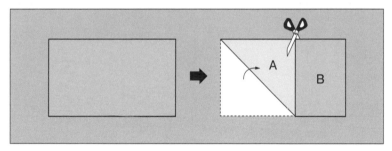

(1) If you open **A**, what shape is it?   (   )

(2) What shape is **B**?   (   )

What is your favorite shape?

Don't forget!

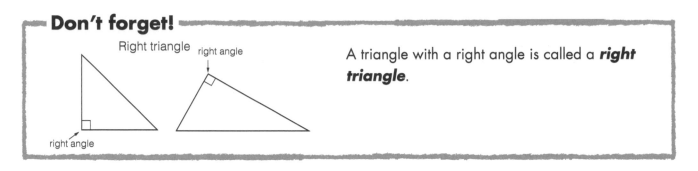

Right triangle right angle

right angle

A triangle with a right angle is called a **right triangle**.

1 Look at the triangles below and circle all the right angles.

20 points for completion

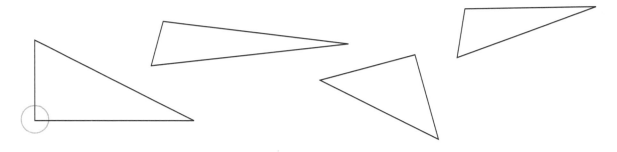

2 Look at the triangles pictured here and write the letter for all the right triangles below.

30 points for completion

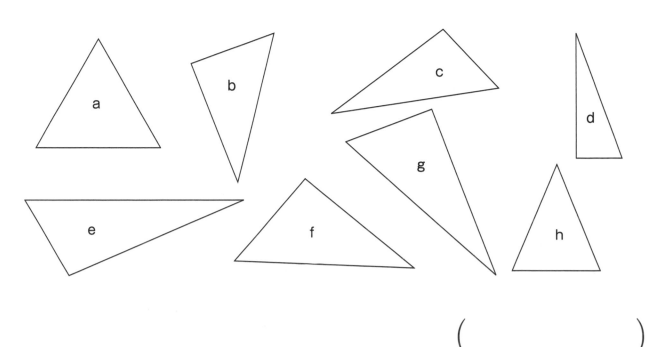

( )

3 These shapes are cut along the dashed lines pictured here. What kind of new shapes did you make? How many?

10 points per question

(1)

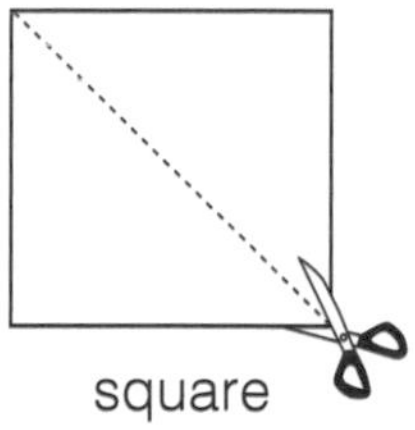

square

name of new shape　　　How many?

(　　　　)　(　　　　)

(2)

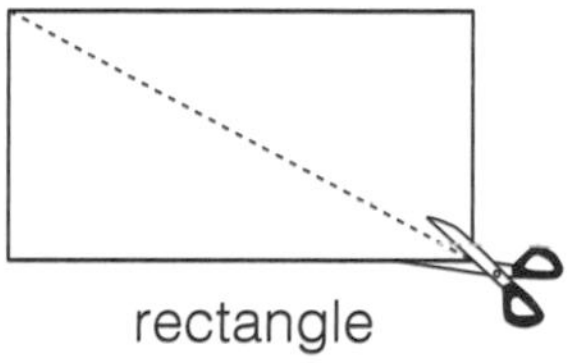

rectangle

(　　　　)　(　　　　)

(3)

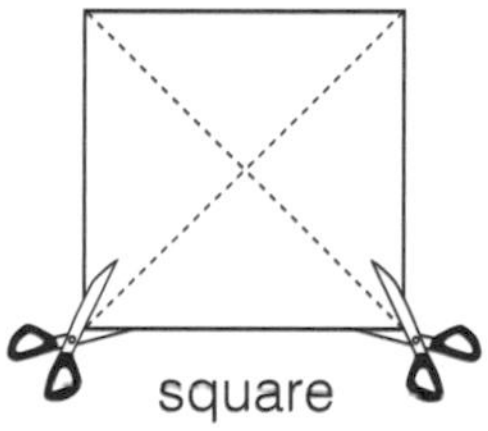

square

(　　　　)　(　　　　)

4 For the following questions, assume you made the shapes below using right triangles.

10 points per question

(1) You used **2** triangles. What shapes did you make?

a

(　　　　)

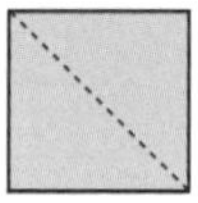

b

(　　　　)

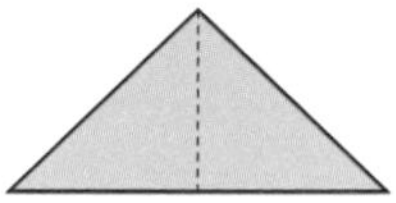

(2) You used **4** triangles. What shapes did you make?

a

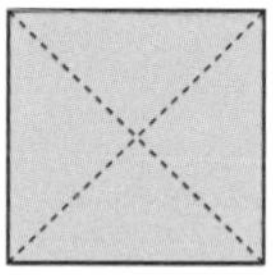

b

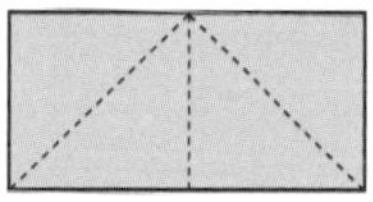

c

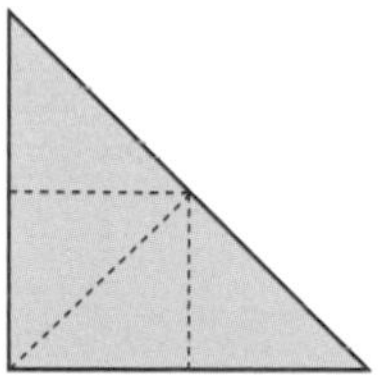

(　　　　)　　(　　　　)　　(　　　　)

Wow, there are triangles everywhere!

Triangles & Quadrilaterals

32

Level ★★

Date   /   /

Name

Score

/100

1 Draw 3 **squares** of different sizes on the grid below.

5 points per square

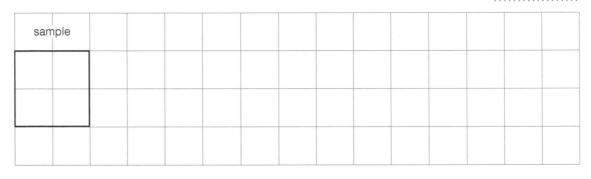

sample

2 Draw 3 **rectangles** of different sizes on the grid below.

5 points per rectangle

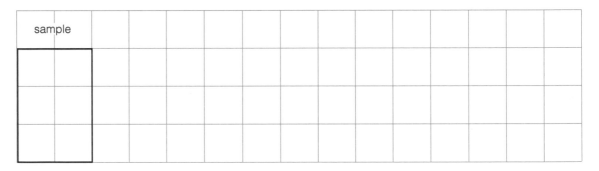

sample

3 Draw 3 **right triangles** of different sizes on the grid below.

5 points per right triangle

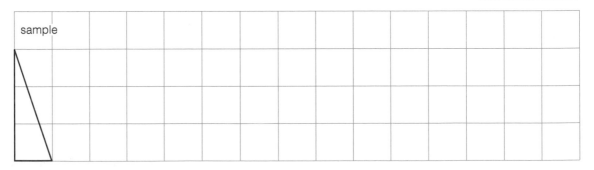

sample

4 Draw the following quadrilaterals on the grid below.

11 points per shape

a   A square with sides that are **4** inches long.

b   A rectangle with sides that are **3** inches and **6** inches long.

c   A rectangle with sides that are **2** inches and **10** inches long.

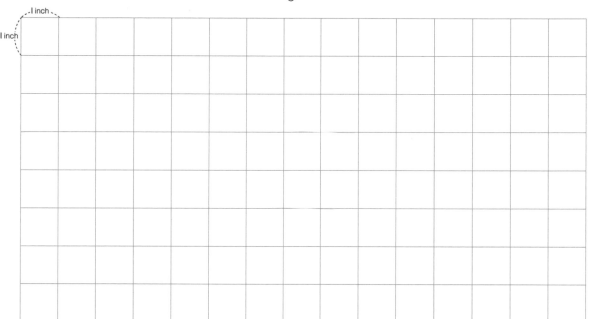

5 Draw the following triangles on the grid below.

11 points per shape

a   Two sides come together at a right angle and are **4** centimeters and **5** centimeters long.

b   Two sides come together at a right angle and are **7** centimeters and **3** centimeters long.

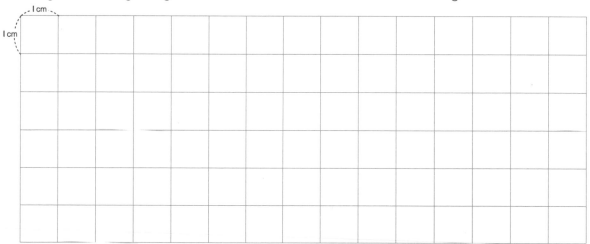

Drawing shapes is fun, right?

Level ★★

Score

Date / /

Name

/100

1 If you cut the boxes pictured here along the bold lines and then opened them up, what would they look like? Write a ✓ under each correct shape on the right.

10 points per question

(1)

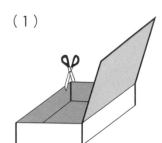

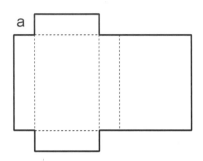

a   b

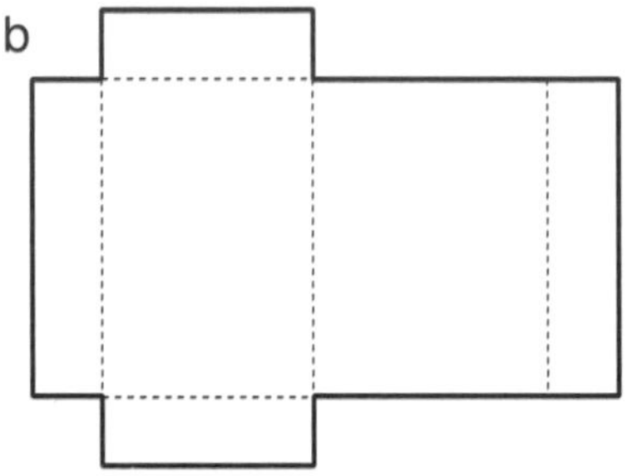

(   )   (   )

(2)

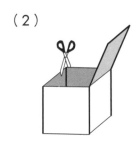

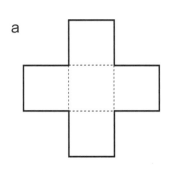

a   b

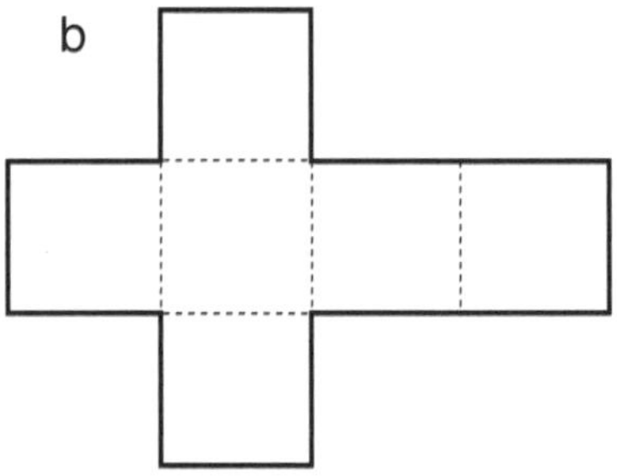

(   )   (   )

(3)

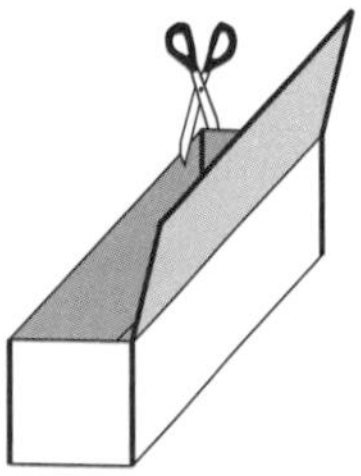

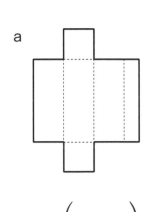

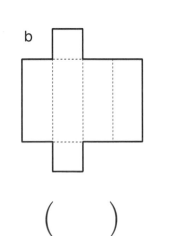

a   b

(   )   (   )

2 The surface of a box is made up of faces. Finish the diagram and then answer the questions below.

3 points per parentheses

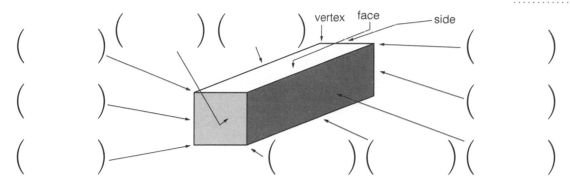

( ) ( ) ( ) vertex face side ( )

( ) ( )

( ) ( ) ( ) ( )

(1) What shapes are the faces on this box?

4 points

( ) and ( )

(2) How many faces does this box have?

( )

2 points

(3) How many sides does this box have?

( )

2 points

(4) How many vertices does this box have?

( )

2 points

3 How many of each type of face do the boxes pictured below have?

10 points per question

(1)

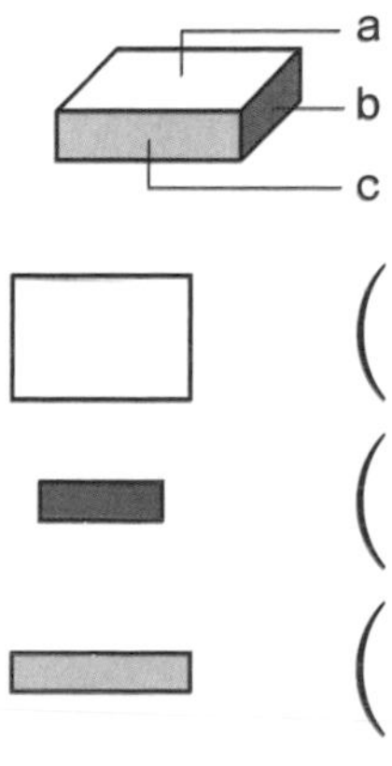

a
b
c

a ▢ ( )

b ▬ ( )

c ▭ ( )

(2)

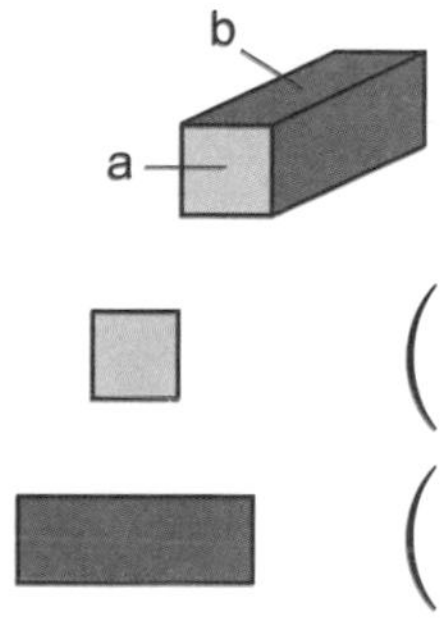

b
a

a ▢ ( )

b ▬ ( )

(3)

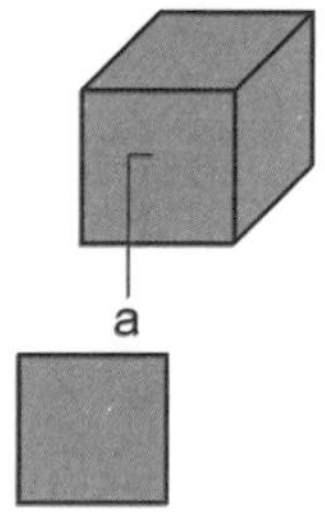

a

a ▢ ( )

You're doing well, keep it up!

34

Boxes

Date

Name

Level
★★

Score

/100

1 If you constructed boxes by folding the figures on the left, what kinds of boxes would you make? Connect the figures on the left to the correct boxes on the right.

40 points per completion

a
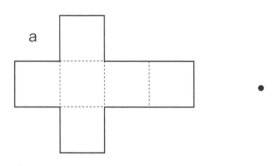

e
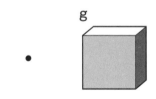

b
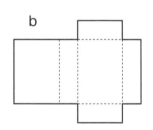

f

c
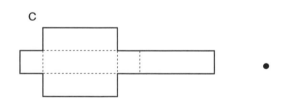

g
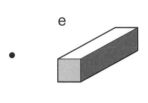

d
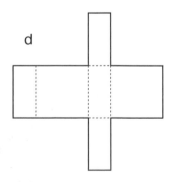

h
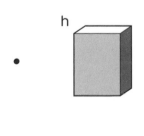

2 A box is constructed from the plan that is pictured here. Answer the questions below.

10 points per question

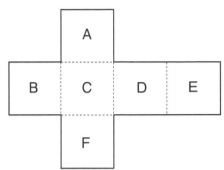

(1) Which face is across from **A**?

( )

(2) Which face is across from **C**?

( )

3 The boxes pictured here are cut along the bold lines and opened. Face A is missing from each picture, though. Please add face A to the figures below.

10 points per question

(1)

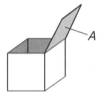

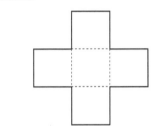

(2)

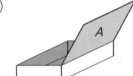

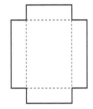

(3)

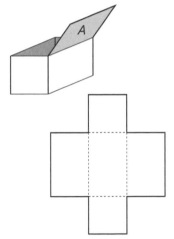

(4)

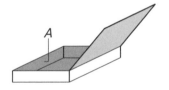

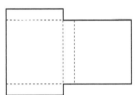

Are you ready to review what you've learned?

Review

35

Date   /   /

Name

Level ☆☆

Score

/100

1 Answer the following questions about the number 37,201,954.

5 points per question

(1)   7 is in the ▢ place.

(2)   0 is in the ▢ place.

(3)   The 3 represents 3 ▢ .

2 Write the appropriate number in each box to complete the sentence.

4 points per question

(1)   The number you get from adding 2 ten-millions, 1 million, 7 hundred-thousands and 4 ten-thousands

is ▢ .

(2)   The number you get from adding 16 ten-thousands is ▢ .

(3)   The number you get from adding 130 thousands is ▢ .

(4)   The number you get from adding 42 ten-thousands and 104 is ▢ .

(5)   The number that is 1,000 less than 1,000,000 is ▢ .

3 What is the weight on each scale below?

5 points per question

(1)

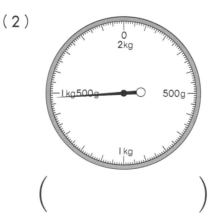

(2)

(3)

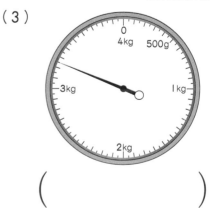

(   )   (   )   (   )

4 How much water is in the containers below?

6 points per question

(1) ^{1 L}

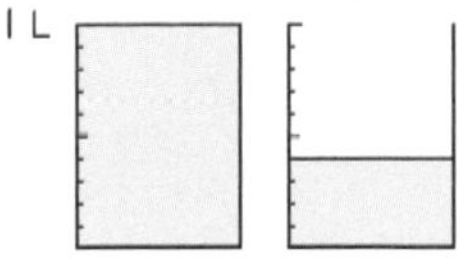

(   )

(2) ^{1 L}

(   )

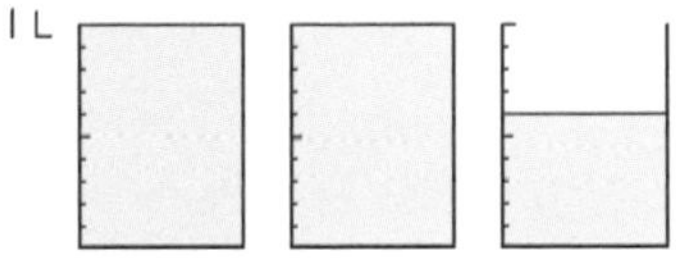

5 Name the shapes below.

6 points per question

(1)

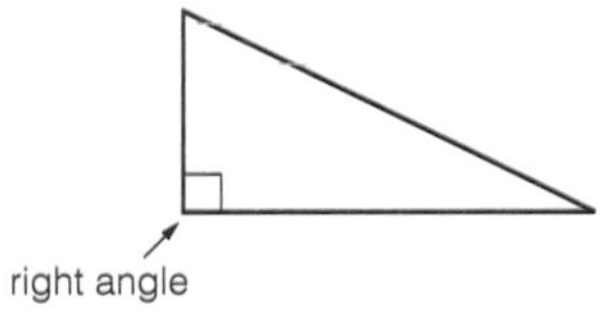

right angle

(   )

(2) right angle   right angle

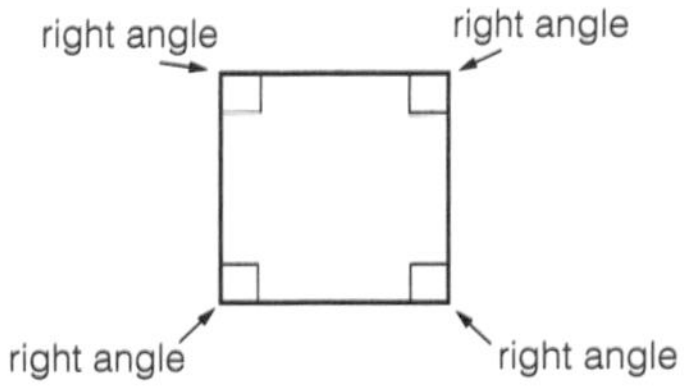

right angle   right angle

(   )

(3) right angle   right angle

right angle   right angle

(   )

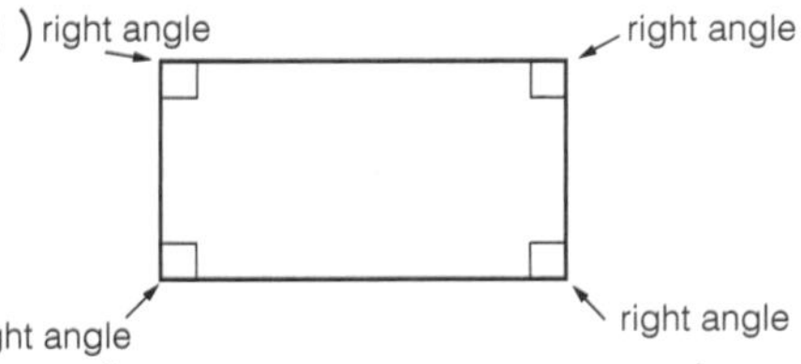

6 Answer the following questions about the box on the right.

4 points per question

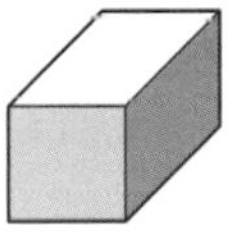

(1) What shapes are the faces on this box?

(   ) and (   )

(2) How many faces does this box have?

(   )

(3) How many sides does this box have?

(   )

(4) How many vertices does this box have?

(   )

(5) Which figure below represents the box if it was opened up?

A

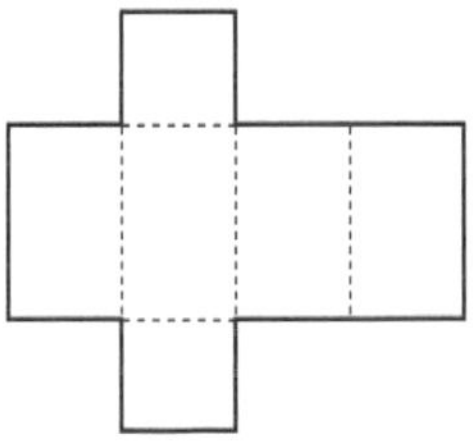

B

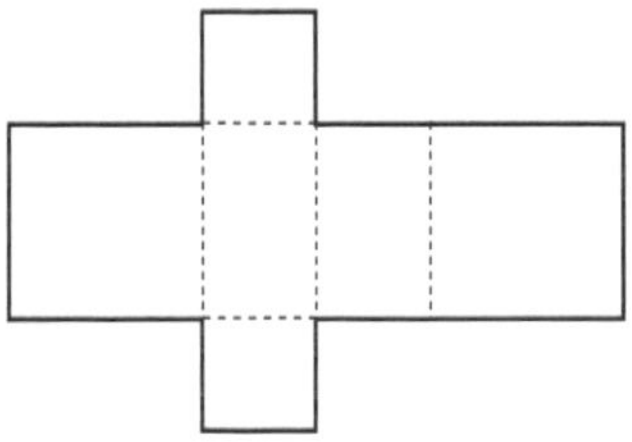

(   )

You're almost there! Well done.

Review

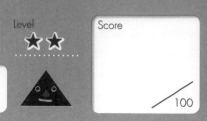

Level
★★

Score
/100

Date   /   /

Name

1 Fill in the missing number in each box on the number lines below.

3 points per box

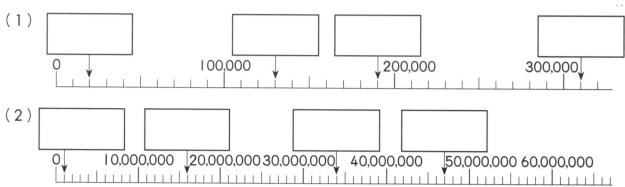

(1)

0   100,000   200,000   300,000

(2)

0   10,000,000 20,000,000 30,000,000 40,000,000 50,000,000 60,000,000

2 Write a ✓ under the larger number.

4 points per question

(1) 1,910,056   1,901,056
     (   )   (   )

(2) 642,900   643,000
     (   )   (   )

3 Complete the table below.

10 points for completion

Base numbers	ten times	hundred times	divided by ten
160			
4,080			

4 Rank the measurements below from longest to shortest.

3 points for completion

1,800 m   1 km 900 m   1,050 m   2 km
(   )   (   )   (   )   (   )

5 Write the time under each clock below.

5 points per question

(1)　　　　　(2)　　　　　(3)　　　　　(4)　　　　　(5)

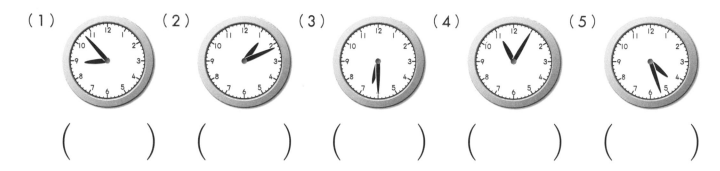

(　　　　　)　(　　　　　)　(　　　　　)　(　　　　　)　(　　　　　)

6 What is the weight on each scale below?

5 points per question

(1)　　　　　　　　　(2)　　　　　　　　　(3)

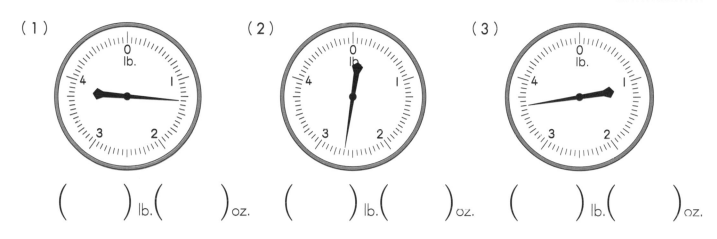

(　　　) lb. (　　　) oz.　(　　　) lb. (　　　) oz.　(　　　) lb. (　　　) oz.

7 How much water is in the containers below?

5 points per question

(1)　　　　　　　　　(2)　　　　　　　　　(3)

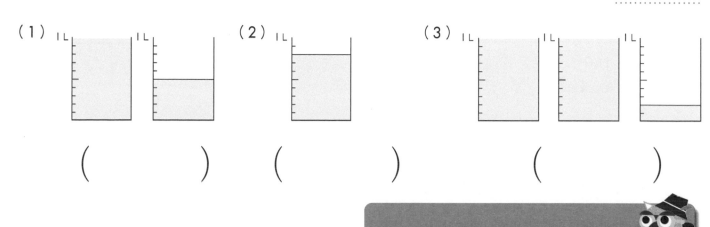

(　　　　　)　　　　　(　　　　　)　　　　　(　　　　　)

Congratulations! You've done a great job.

(1) Review
pp 2, 3

1
(1) 1,465 (2) 5,743 (3) 792
(4) 801 (5) 6,340 (6) 10,000
(7) 230 (8) 9,000 (9) 8,600
(10) 2,530 (11) 9,008

2
(1) 6:30 (2) 9:15
(3) 10:50 (4) 4:25

3
(1) 3 in. (2) 5 in.

4
(1) $1.38 (2) $1.61 (3) $5.07
(4) $10.11 (5) $10.39

(2) Review
pp 4, 5

1
(1) (From the left) 9,900, 9,940, 9,960, 9,990
(2) 4,996, 4,998, 5,000, 5,003

2
(1) 6,001 (2) 9,999 (3) 5,011
(4) 5,009 (5) 9,000 (6) 7,899

3
(1) 9,700 (2) 5,430
(3) 1,061 (4) 6,767
(5) 4,333 (6) 8,108

4
(1) (2) (3)

5 Triangles···ⓑ, ⓕ, ⓙ
Quadrilaterals···ⓐ, ⓒ, ⓖ, ⓘ

6
(1) 3 in. (2) 4 in. (3) 2 in.

(3) Large Numbers
pp 6, 7

1
(1) 9,000 (2) 10,000
(3) 11,000 (4) 10,000
(5) 20,000 (6) 22,000

2
(1) 25,143 (2) 36,215 (3) 40,234
(4) 51,006

3
(1) 79,462 (2) 52,801 (3) 45,000
(4) 90,800 (5) 30,070

(4) Large Numbers
pp 8, 9

1
(1) 10,000 (2) 100,000
(3) 1,000,000 (4) 10,000,000

2
(1) (From the left) ten-millions, millions, hundred-thousands
(2) ten-thousands, hundred-thousands, millions, ten-millions
(3) millions
(4) ten-millions
(5) hundred-thousands

3
(1) 43,570,000 (2) 68,001,020
(3) 21,760,000 (4) 85,010,000
(5) 9,300,000

4
(1) 7 (2) 17 (3) 170 (4) 170
(5) 1,700 (6) 35 (7) 150,000

(5) Large Numbers
pp 10, 11

1
(1) (From the left) 6,000, 15,000, 24,000, 31,000
(2) 10,000, 90,000, 190,000, 250,000
(3) 100,000, 110,000, 130,000
(4) 19,100, 20,100, 21,600
(5) 99,100, 100,300, 101,000

2
(1) 100,000 (2) 100,100
(3) 99,900

3
(1) 40,000 (2) 300,000
(3) 290,000 (4) 38,500

4
(1) ⌈ 876,539 867,539 875,639 ⌉
    (1)   (3)   (2)
(2) ⌈ 99,999 100,200 100,120 ⌉
    (3)   (1)   (2)

5
(1) > (2) < (3) > (4) <
(5) > (6) < (7) > (8) >
(9) <

6 **Large Numbers** pp 12,13

1 (1) 100 (2) 1,000 (3) 10,000
(4) 1,100 (5) 11,100
(6) 520 (7) 5,200

2

Base number	38	47	60	500
ten times	380	470	600	5,000
hundred times	3,800	4,700	6,000	50,000
Base number	945	106	120	790
ten times	9,450	1,060	1,200	7,900
hundred times	94,500	10,600	12,000	79,000

3 (1) 520 (2) 52

4 (1) 7 (2) 8 (3) 40 (4) 300
(5) 19 (6) 61 (7) 920 (8) 902

7 **Fractions** pp 14,15

1 (1) 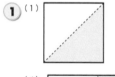 (2) (3)

(4)  (5)

(Coloring either side is correct.)

2 (1) $\frac{1}{2}$ (2) $\frac{1}{3}$ (3) $\frac{1}{4}$ (4) $\frac{1}{4}$ (5) $\frac{1}{5}$ (6) $\frac{1}{6}$
(7) $\frac{1}{6}$

3 (1) $\frac{1}{2}$ (2) $\frac{1}{3}$ (3) $\frac{1}{4}$ (4) $\frac{1}{6}$

4 (1) $\frac{1}{2}$ (2) $\frac{1}{3}$ (3) $\frac{1}{4}$ (4) $\frac{1}{5}$

8 **Fractions** pp 16,17

1 (1) $\frac{3}{5}$ (2) $\frac{4}{5}$ (3) $\frac{5}{6}$ (4) $\frac{2}{7}$ (5) $\frac{4}{7}$ (6) $\frac{3}{6}$
(7) $\frac{5}{8}$ (8) $\frac{2}{9}$

2 (1) $\frac{1}{4}$ (2) $\frac{2}{4}$ (3) $\frac{3}{4}$ (4) $\frac{2}{5}$ (5) $\frac{4}{5}$

3 (1) $\frac{2}{6}$ (2) $\frac{4}{6}$ (3) $\frac{5}{6}$ (4) $\frac{3}{10}$
(5) $\frac{5}{9}$ (6) $\frac{5}{12}$ (7) $\frac{7}{12}$

9 **Length** pp 18,19

1 A $\frac{1}{2}$ B $\frac{1}{4}$ C $\frac{1}{8}$ D $\frac{3}{4}$
E $\frac{3}{8}$ F $\frac{1}{16}$ G $\frac{5}{16}$ H $\frac{7}{16}$

2 (1) 1 in. (2) $1\frac{1}{2}$ in. (3) $1\frac{1}{4}$ in. (4) $1\frac{3}{4}$ in.
(5) $1\frac{1}{8}$ in. (6) $1\frac{5}{8}$ in. (7) $1\frac{9}{16}$ in. (8) $1\frac{11}{16}$ in.

3 (1) 2 in. (2) $1\frac{3}{4}$ in. (3) $1\frac{7}{16}$ in. (4) $1\frac{7}{8}$ in.
(5) 1 in. (6) $1\frac{15}{16}$ in.

10 **Length** pp 20,21

1 (1) 5,280 (11) 1
(2) 5,285 (12) 1, 10
(3) 5,330 (13) 1, 220
(4) 5,780 (14) 1, 720
(5) 5,820 (15) 1, 1,720
(6) 10,560 (16) 1, 3,720
(7) 10,630 (17) 2
(8) 10,640 (18) 2, 440
(9) 10,860 (19) 2, 2,440
(10) 15,860 (20) 3

2 (1) 900 ft. (2) 720 ft.
(3) 1 mi. (4) 1 mi. 300 ft.
(5) 1 mi. 100 ft. (6) 1 mi. 950 ft.
(7) 6,000 ft. (8) 6,500 ft.
(9) 1 mi. 40 ft. (10) 5,300 ft.

3 (1) (From the left) 3, 1, 2, 4
(2) 2, 1, 4, 3

11 **Length** pp 22,23

1 (1) 1 mm (2) 2 mm (3) 3 mm (4) 5 mm
(5) 4 mm (6) 6 mm (7) 8 mm (8) 9 mm

2 (1) 10 mm (2) 11 mm (3) 13 mm (4) 15 mm
(5) 12 mm (6) 16 mm (7) 17 mm (8) 19 mm

3 (From the left) 10 mm, 30 mm, 50 mm, 70 mm, 90 mm, 100 mm

4 (1) 20 mm (2) 21 mm (3) 25 mm
(4) 27 mm (5) 30 mm (6) 38 mm

5 (1) 9 mm (2) 22 mm

12 Length
pp 24, 25

1 (1) (From the left) 1 cm 5 mm, 5 cm 1 mm,
9 cm 5 mm

(2) 4 cm 3 mm, 8 cm 7 mm, 12 cm 3 mm

2 (1) 7 cm 6 mm
(2) 12 cm 9 mm

3 (1) 5 cm 5 mm (2) 7 cm 6 mm (3) 8 cm 4 mm

4

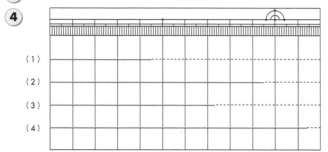

(1)
(2)
(3)
(4)

13 Length
pp 26, 27

1 (1) 1,000 m (11) 1 km
(2) 1,005 m (12) 1 km 400 m
(3) 1,050 m (13) 1 km 60 m
(4) 1,500 m (14) 1 km 850 m
(5) 1,550 m (15) 2 km 300 m
(6) 2,000 m (16) 2 km 50 m
(7) 2,080 m (17) 2 km 560 m
(8) 2,400 m (18) 3 km
(9) 2,650 m (19) 3 km 200 m
(10) 3,180 m (20) 3 km 650 m

2 (1) 800 m (2) 650 m
(3) 1 km (4) 1 km 400 m
(5) 1 km 100 m (6) 1 km 650 m
(7) 1 km 200 m (8) 1,380 m
(9) 1,600 m (10) 2 km 500 m

3 (1)

1 km	1,100 m	1 km 10 m	990 m
(3)	(1)	(2)	(4)

(2)

1,100 m	2 km	1 km 790 m	2,050 m
(3)	(2)	(4)	(1)

14 Area
pp 28, 29

1 (1) Ⓐ 8 Ⓑ 4 Ⓒ 4
(2) Ⓐ

2 (1) 1 in.² (2) 2 in.² (3) 2 in.²
(4) 3 in.² (5) 3 in.² (6) 4 in.²

3 (1) Ⓐ 4 Ⓑ 6 Ⓒ 7
(2) Ⓒ

4 (1) 1 cm² (2) 2 cm² (3) 2 cm²
(4) 3 cm² (5) 3 cm² (6) 4 cm²

15 Weight
pp 30, 31

1 (1) 1 (2) 2 (3) 5 (4) 10
(5) 1 (6) 1, 8 (7) 1, 15 (8) 2, 10

2 (1) 16 (2) 17 (3) 19 (4) 21
(5) 26 (6) 31 (7) 32 (8) 34
(9) 37 (10) 52 (11) 1 (12) 1, 1
(13) 1, 2 (14) 1, 4 (15) 1, 10 (16) 1, 11
(17) 1, 14 (18) 2, 8 (19) 2, 10 (20) 2, 13

16 Weight
pp 32, 33

1 (1) 200 g (2) 250 g (3) 350 g
(4) 750 g (5) 950 g (6) 70 g
(7) 180 g (8) 330 g (9) 540 g

2 (1) 10 g (2) 50 g (3) 90 g
(4) 120 g (5) 170 g (6) 230 g
(7) 340 g (8) 460 g (9) 670 g
(10) 980 g (11) 1 kg

17 Weight
pp 34, 35

1 (1) 1,000 g (11) 1 kg
(2) 1,050 g (12) 1 kg 200 g
(3) 1,500 g (13) 1 kg 60 g
(4) 2,000 g (14) 2 kg 100 g
(5) 2,010 g (15) 2 kg 80 g
(6) 2,100 g (16) 2 kg 500 g
(7) 2,800 g (17) 3 kg
(8) 3,000 g (18) 3 kg 600 g
(9) 3,040 g (19) 4 kg
(10) 3,400 g (20) 4 kg 80 g

2 (1) 800 g (2) 550 g
(3) 1 kg (4) 1 kg 300 g
(5) 1 kg 100 g (6) 1,100 g
(7) 1 kg 500 g (8) 2,500 g
(9) 2,080 g (10) 4,600 g

3 (1)

1 kg	1,010 g	1 kg 100 g	990 g
(3)	(2)	(1)	(4)

(2)

3 kg 700 g	4 kg	3,800 g	3,090 g
(3)	(1)	(2)	(4)

18 Weight
pp 36,37

1 (1) 1 kg 200 g (2) 1 kg 700 g
(3) 1 kg 900 g (4) 1 kg 350 g
(5) 1 kg 550 g (6) 1 kg 60 g
(7) 1 kg 110 g (8) 1 kg 730 g

2 (1) 100 g (2) 800 g
(3) 1 kg 200 g (4) 2 kg 350 g
(5) 3 kg 300 g (6) 3 kg 850 g

19 Weight
pp 38,39

1 (1) 500 mg (2) 850 mg

2 (1) 1,000 (6) 1
(2) 2,000 (7) 3
(3) 4,000 (8) 5
(4) 6,000 (9) 7
(5) 10,000 (10) 9

3 (1) 800 mg (2) 350 mg
(3) 1 g (4) 2,010 mg
(5) 1,101 mg (6) 3,010 mg

4 (1)

1 g	1,001 mg	995 mg	1,100 mg
(3)	(2)	(4)	(1)

(2)

1 kg	1 mg	1 g	10,000 mg
(1)	(4)	(3)	(2)

(3)

2,500 mg	3 g	2,010 mg	2 g
(2)	(1)	(3)	(4)

(4)

10 mg	1 mg	1 g	100 mg
(3)	(4)	(1)	(2)

20 Capacity
pp 40,41

1 (1) (From the left) 2, 3, 1
(2) 3, 1, 2
(3) 1, 3, 2
(4) 1, 2, 3

2 Bob

3 kettle, 2

4 (From the top) 2, 1, 3

21 Capacity
pp 42,43

1 (1) 1 pt. (2) 2 pt. (3) 3 pt.
(4) 5 pt.

2 (1) 1 qt. (2) 2 qt. (3) 4 qt. (4) 6 qt.

3 (1) 2 (11) 1
(2) 3 (12) 2
(3) 4 (13) 2, 1
(4) 6 (14) 4
(5) 7 (15) 5, 1
(6) 9 (16) 7
(7) 10 (17) 7, 1
(8) 12 (18) 8
(9) 13 (19) 8, 1
(10) 15 (20) 10

22 Capacity
pp 44,45

1 (1) 1 gal. (2) 2 gal. (3) 3 gal. (4) 4 gal.

2 (1) 4 (5) 1
(2) 5 (6) 1, 2
(3) 7 (7) 2, 1
(4) 8 (8) 2, 2

3 (1) 1 gal. 2 qt., 6 qt.
(2) 1 gal. 1 qt., 5 qt.
(3) 2 gal., 8 qt.
(4) 2 gal. 3 qt., 11 qt.
(5) 2 qt. 1 pt., 5 pt.
(6) 3 qt., 6 pt.
(7) 4 qt., 8 pt.
(8) 1 gal., 4 qt., 8 pt.

(23) Capacity
pp 46,47

(1) (1) 1 L   (2) 2 L   (3) 5 L   (4) 7 L

(2) (1) 1,000   (5) 1
(2) 2,000   (6) 3
(3) 4,000   (7) 9
(4) 8,000   (8) 11

(3) (1) 100 mL   (2) 200 mL   (3) 400 mL
(4) 600 mL   (5) 700 mL   (6) 900 mL
(7) 1,000 mL (1 L)

(4) (1) 500 mL   (2) 900 mL   (3) 1 L 100 mL
(4) 1 L 300 mL

(24) Telling Time
pp 48,49

(1) (1) 8:00   (2) 8:01   (3) 8:02   (4) 8:03
(5) 8:04   (6) 8:05

(2) (1) 8:02   (2) 8:05   (3) 8:01   (4) 8:04

(3) (1) 8:05   (2) 8:06   (3) 8:07   (4) 8:08
(5) 8:09   (6) 8:10

(4) (1) 8:08   (2) 8:06   (3) 8:10   (4) 8:09

(25) Telling Time
pp 50,51

(1) (1) 8:05   (2) 6:05   (3) 8:04   (4) 10:04

(2) (1) 2:01   (2) 7:01   (3) 9:03   (4) 6:03
(5) 4:02   (6) 12:04

(3) (1) 8:15   (2) 8:16   (3) 8:17   (4) 8:18
(5) 8:19   (6) 8:20

(4) (1) 8:17   (2) 8:20   (3) 8:21   (4) 8:16

(26) Telling Time
pp 52,53

(1) (1) 8:05   (2) 8:10   (3) 8:15   (4) 8:20
(5) 8:25   (6) 8:30   (7) 8:35   (8) 8:40
(9) 8:45   (10) 8:50   (11) 8:55   (12) 8:25

(2) (1) 8:10   (2) 11:10   (3) 8:08   (4) 5:08

(3) (1) 9:06   (2) 6:06   (3) 10:07   (4) 2:07
(5) 1:09   (6) 12:08

(27) Telling Time
pp 54,55

(1) (1) 8:10   (2) 8:11   (3) 8:12   (4) 8:13
(5) 8:14   (6) 8:15

(2) (1) 8:11   (2) 8:15   (3) 8:13   (4) 8:14

(3) (1) 8:15   (2) 8:16   (3) 8:17   (4) 8:18
(5) 8:19   (6) 8:20

(4) (1) 8:17   (2) 8:20   (3) 8:21   (4) 8:16

(28) Telling Time
pp 56,57

(1) (1) 9:17   (2) 10:17   (3) 2:17   (4) 3:17

(2) (1) 1:22   (2) 4:22   (3) 5:49   (4) 2:49
(5) 11:28   (6) 5:28

(3) (1) 3:47   (2) 1:38   (3) 4:53   (4) 4:52

(4) (1) 4:42   (2) 9:21   (3) 10:54   (4) 11:57
(5) 9:47   (6) 6:33

(29) Triangles & Quadrilaterals
pp 58,59

(1)

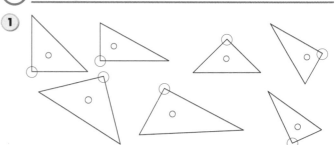

(2) (1) ✓   (2) ✕   (3) ✓

(3) (1) 4   (2) 4   (3) 2 pairs

(4) a, b, e, i, k

(30) Triangles & Quadrilaterals
pp 60,61

(1) (1) 4   (2) same (yes)

(2) c, f, j

(3) (1) b, d, f   (2) c, h, i
(3) a, e, g

(4) (1) Square   (2) Rectangle

(31) Triangles & Quadrilaterals   pp 62,63

1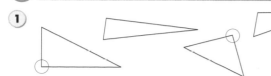

2 b, d, f, g

3 (1) Right triangle, 2   (2) Right triangle, 2
(3) Right triangle, 4

4 (1) a Square   b Right triangle
(2) a Square   b Rectangle   c Right triangle

(32) Triangles & Quadrilaterals   pp 64,65

1 ⟨Answer Example⟩
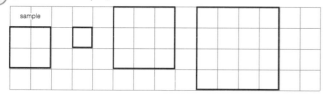

2 ⟨Answer Example⟩

3 ⟨Answer Example⟩
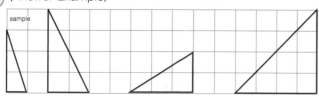

4 ⟨Answer Example⟩
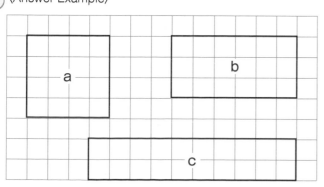

5 ⟨Answer Example⟩
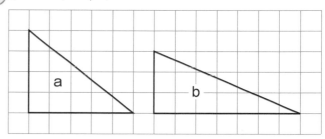

(33) Boxes   pp 66,67

1 (1) a   (2) b   (3) b

2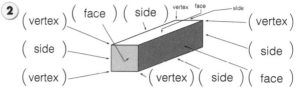
(vertex) (face) (side)   (vertex)
(side)   (side)
(vertex)   (vertex) (side) (face)

(1) Rectangle, Square
(2) 6   (3) 12   (4) 8

3 (1) a : 2   b : 2   c : 2
(2) a : 2   a : 4   (3) a : 6

(34) Boxes   pp 68,69

1 a—f   b—h   c—e   d—g

2 (1) F   (2) E

3 (1)
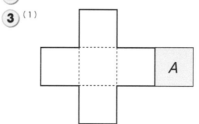

(2)
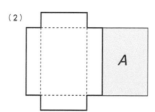

79

(3)

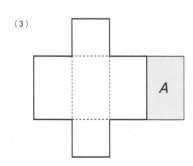

(4)

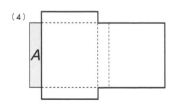

③⑤ Review
pp 70,71

1 (1) millions (2) ten-thousands (3) ten-millions

2 (1) 21,740,000 (2) 160,000
  (3) 130,000 (4) 420,104
  (5) 999,000

3 (1) 630 g (2) 1 kg 480 g (3) 3 kg 250 g

4 (1) 1 L 400 mL (2) 2 L 600 mL

5 (1) Right triangle (2) Square (3) Rectangle

6 (1) Rectangle, Square (2) 6 (3) 12
  (4) 8 (5) A

③⑥ Review
pp 72,73

1 (1) (From the left) 20,000, 130,000, 190,000, 310,000
  (2) 1,000,000, 16,000,000, 34,000,000, 47,000,000

2 (1) 1,910,056 (2) 643,000

3

Base number	ten times	hundred times	divided by ten
160	1,600	16,000	16
4,080	40,800	408,000	408

4 (1) 1,800 m   1 km 900 m   1,050 m   2 km
     (3)   (2)   (4)   (1)

5 (1) 8:53 (2) 1:11 (3) 6:30 (4) 11:05
  (5) 4:27

6 (1) 1 lb. 5 oz. (2) 2 lb. 10 oz. (3) 3 lb. 10 oz.

7 (1) 1 L 500 mL (2) 800 mL (3) 2 L 200 mL